Contemporary Issues in Wireless Communications

Contemporary Issues in Wireless Communications

Editor

Fredrik Jonasen

scitus
academics

Contemporary Issues in Wireless Communications
Edited by **Fredrik Jonasen**

ISBN: 978-1-68117-285-9
Library of Congress Control Number: 2016937461

SCITUS Academics LLC,
www.scitusacademics.com
Box No. 4766, 616 Corporate Way,
Suite 2, Valley Cottage,
NY 10989

Notice

Printed in United States of America on Acid Free Paper

Preface

In 21st Century, Wireless communications have a robust effect on enlightening the quality of life. Smart phones industry is now well thought-out one of the most gorgeous fields, so advanced research is conducted in order to improve the quality of service in wireless communication environments. Wireless communication is among technology's biggest contributions to mankind. Wireless communication involves the transmission of information over a distance without help of wires, cables or any other forms of electrical conductors. The transmitted distance can be anywhere between a few meters such as a television's remote control and thousands of kilometres such as radio communication. Some of the devices used for wireless communication are cordless telephones, mobiles, GPS units, wireless computer parts, and satellite television. The growth of wireless network has enabled us to use personal devices anywhere and anytime. This has helped mankind to improve in every field of life but this has led many threats as well. Wireless network has led to many security threats to mankind. It is very easy for the hackers to grab the wireless signals that are spread in the air. It is very important to secure the wireless network so that the information cannot be exploited by the unauthorized users. This also increases the risk to lose information. Strong security protocols must be created to secure the wireless signals like WPA and WPA2. Another way to secure the wireless network is to have wireless intrusion prevention system. Many design challenges such as power consumption, quality of service, low cost, high data rate and small size are being preserved day by day. This book entitled Contemporary Issues in Wireless Communications provide highlights of the contemporary research in the field of wireless communications.

Table of Contents

CHAPTER 1

A review of communication-oriented optical wireless systems

Deva K Borah[1], Anthony C Boucouvalas[2], Christopher C Davis[3], Steve Hranilovic[4] and Konstantinos Yiannopoulos[5*]

[1]Klipsch School of Electrical and Computer Engineering, New Mexico State University, Las Cruces, NM, USA

[2]Department of Telecommunications Science and Technology, University of Peloponnese, Tripoli, Greece

[3]Department of Electrical and Computer Engineering, University of Maryland, College Park, MD, USA

[4]Department of Electrical and Computer Engineering, McMaster University, Hamilton, ON, Canada

[5]Department of Telecommunications Science and Technology, University of Peloponnese, Tripoli, Greece

ABSTRACT

This article presents an overview of optical wireless (OW) communication systems that operate both in the short- (personal and indoor systems) and the long-range (outdoor and hybrid) regimes. Each of these areas is discussed in terms of (a) key requirements, (b) their application framework, (c) major impairments and applicable mitigation techniques, and (d) current and/or future trends. Personal communication systems are discussed within the context of point-to-point ultra-high speed data transfer. The most relevant application framework and related standards are presented, including the next generation Giga-IR standard that extends personal communication speeds to over 1 Gb/s.

As far as indoor systems are concerned, emphasis is given on modeling the dispersive nature of indoor OW channels, on the limitations that dispersion imposes on user mobility and dispersion mitigation techniques. Visible light communication systems, which provide both illumination and communication over visible or hybrid visible/infrared LEDs, are presented as the most important representative of future indoor OW systems. The discussion on outdoor systems focuses on the impact of atmospheric effects on the optical channel and associated mitigation techniques that extend the realizable link lengths and transfer rates. Currently, outdoor OW is commercially available at 10 Gb/s Ethernet speeds for Metro networks and Local-Area-Network interconnections and speeds are expected to increase as faster and more reliable optical components become available. This article concludes with hybrid optical wireless/radio-frequency (OW/RF) systems that employ an additional RF link to improve the overall system reliability. Emphasis is given on cooperation techniques between the reliable RF subsystem and the broadband OW system.

KEYWORDS

optical wireless personal communication systems; indoor systems; outdoor systems; hybrid OW/RF systems

1. INTRODUCTION

It has been more than 30 years since optical wireless (OW) was proposed as an alternative broadband technology for wireless data transmission applications [1]. The underlying concept of OW is very simple: utilize optical beams to carry data through the atmosphere or vacuum. As a result, OW link architectures are very similar to optical fiber communication point-to-point links, with the exception that no optical fibers are deployed as a transmission medium. OW is also very similar to RF wireless, but radio waves are replaced with light and antennas with free-space optical transceivers. Despite this superficial resemblance between OW and RF links, OW exhibits several appealing attributes when compared to RF. OW links are inherently broadband and optical frequencies in the infrared and visible spectrum are neither regulated nor licensed. Optical components are also cheaper and consume less electrical power than high-speed RF components. Finally, OW links do not suffer from multipath fading and have much less potential for interference with RF-sensitive electronic systems. These advantages do not, however, imply that OW is a universal replacement for RF communications. The application of OW systems is limited when considering area coverage and user mobility, where RF technologies prove invaluable. In addition, OW systems operate under strict eye safety regulations, while at the same time incoherent OW receivers present lower sensitivity than their RF counterparts because of their photo-electric conversion mechanisms and the impact of ambient light noise sources. Table 1 summarizes the main differences between OW and RF systems.

Table 1 Comparison between RF and OW systems

	OW	RF
Bandwidth	Not regulated	Licensed
Available line rates	< 10 Gb/s	< 1.25 Gb/s
Path losses	High	High
Multipath fading	No (large collector area)	Yes
Multipath distortion	Only in diffuse indoor systems	Yes
Noise sources	Ambient light	Interference from other users, electrical noise
Detection type	Incoherent	Coherent/Incoherent
SNR	Depends on optical signal power	Depends on RF signal amplitude
Receiver sensitivity	Low	High
Eye safety	Required	N/A
Electromagnetic compatibility	Yes	Conditional

To better understand the place of OW systems in the wireless world, Figure 1 summarizes state-of-the art commercial RF and OW technologies, as well as technologies under standardization by major bodies including IEEE, 3GPP, Bluetooth and IrDA. Technologies are presented with respect to their area of coverage, ranging from a few centimeters in personal communications to over 1 km in outdoor communications, and the data rates they attain, including low rate legacy links under 1 Mb/s (Bluetooth and older IrDA systems). Clearly, contemporary OW links provide channel rates up to 10 Gb/s, which directly compare to the ones of optical fibers. At the same time, commercial OW links operate at link distances that are challenging to attain in RF (3G/4G) and millimeter-wave (60 GHz) broadband communications.

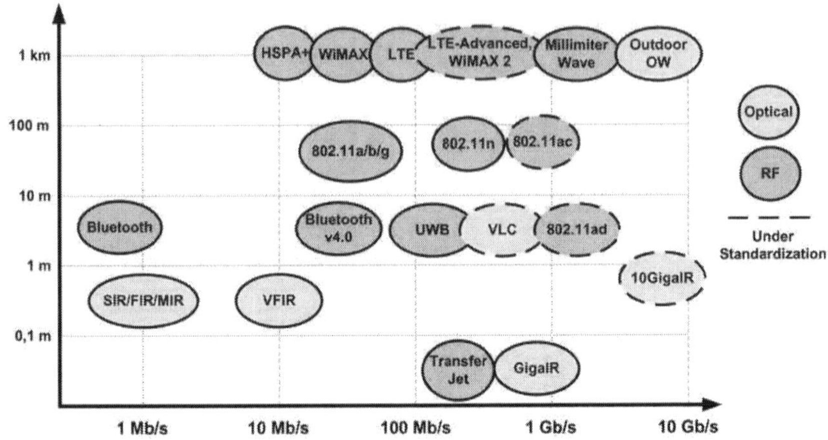

Figure 1 RF and OW technologies as defined in standards and deployed in commercial products.

OW is a unique technology that provides an attractive alternative in niche application areas, complementing fiber-optic and RF wireless solutions when they are either too costly to deploy, create undesirable interference, or are not feasible at all. Figure 2 illustrates some of the application areas in which OW has been successfully applied. Two mainstream application areas of OW are last-mile broadband access and office interconnection; both are the business objectives of a number of component and system manufacturers. In such applications, state-of-the-art OW systems support 10 Gb/s Ethernet, which equals the bandwidth provided by metro fiber optic systems and is significantly higher than the 1.25 Gb/s Ethernet provided by competing RF wireless systems that operate in the 60 GHz frequency range. At the same time the deployment cost of OW systems is significantly lower than that of fiber optics, which can easily reach $1M/mile in urban areas. OW installation only requires the alignment of two free-space optical transceivers rather than digging trenches and repairing roads.

Another major application area of OW is in personal communication systems. The rapid progress in optical technology over the past 30 years, driven mainly by fiber-optic systems and display technologies, has enabled the mass production of high volume, low cost, and fast optical components that are suitable for short range OW. During the 1990s OW emerged as a candidate technology for data transmission in personal communications systems using protocols developed by the Infrared Data Association [2]. The result of this study was the standardization and commercialization of OW ports that have been extremely popular since the late 1990s and may be found on all kinds of mobile phones and portable computers. The current state-of-the-art in personal communications is Gigabit Infrared (Giga-IR) that operates at link speeds of 512 Mb/s and 1.024 Gb/s. This short-range interconnection is among the highest speed wireless interconnection media available and is easily integrated into portable and consumer devices. At the Expo Comm Wireless Japan in 2008 and 2009, KDDI Corporation demonstrated Giga-IR, which is envisioned as an interconnection medium for future cellular phones.

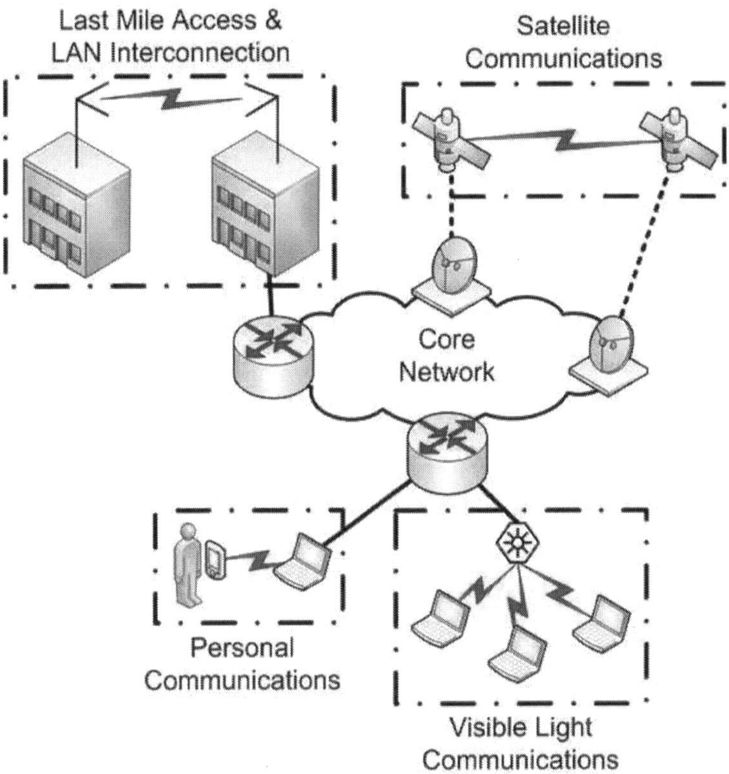

Figure 2 Application areas of optical wireless systems.

Figure 2 also illustrates two additional application areas of OW: indoor and ultra-long-haul systems. OW was initially proposed as a broadband transmission technology for short range diffuse indoor wireless systems [1]. Indoor OW systems are currently being revisited within the framework of Visible Light Communication (VLC) Systems [3] that aim to combine lighting and communications over commercially available visible light LEDs. With respect to ultra-long-haul systems, major technological advances in the fields of adaptive optics and beam acquisition/tracking, which are functions of critical importance in ultra-long-haul outdoor OW links, have enabled the deployment of tractable broadband outdoor OW communication systems. At the time of writing of this article, numerous ultra-long-haul outdoor OW systems have been proposed and demonstrated in a wide range of applications, including terrestrial, earth-to-satellite, earth-to-high altitude platform, intersatellite, and interplanetary communication links [4].

OW is a standalone technology, differentiated from fiber optics and RF wireless as far as usage scenarios are concerned. OW fits perfectly well into the broad scenario of telecommunication networks as a complementary technology that provides communication solutions tailored to diverse application areas. This article aims to present progress in OW systems, underlying principles that lay the basis of OW, and some of the newest trends that appear to be of practical importance in the future. To facilitate the presentation, OW systems have been categorized in four distinct areas: personal communication systems, indoor communication systems, outdoor communication systems, and hybrid OW/RF systems. Personal communication systems are presented in Section 2. The discussion focuses on the most relevant user requirements and standards (IrDA and IEEE), and also includes the Giga-IR extensions that are the latest trend in short range OW systems. Section 3 presents indoor communication systems, with emphasis given to suitable channel models for the diffuse OW link and techniques for achieving spatial coverage. Section 3 also details the latest trend in indoor OW, Visible-Light Communication Systems. Outdoor OW systems are discussed in Section 4. Emphasis is given to the negative impact that the atmosphere has on optical beams, as well as suitable mitigation techniques that largely alleviate this impact. The latest advances in the most challenging application area of outdoor OW systems, satellite, and high-altitude-platform communications, are also presented. This article concludes with Section 5, which presents outdoor hybrid OW/RF systems. The discussion revolves around the cooperation schemes between the OW and RF links that achieve increased system reliability and capacity availability.

2. PERSONAL COMMUNICATION SYSTEMS

Short-range OW has attracted considerable attention for personal communication systems over the past decade, owing to the fact that OW systems offer a viable, low cost and complexity, high bandwidth solution to user-terminal connectivity. OW provides considerable unlicensed (free) bandwidth in the infrared spectrum, the use of which results in significant cost savings to equipment manufacturers since they do not have to pay an extra fee for using this spectrum and their transceiver designs do not have to strictly conform to spectral masks. Additionally, OW is preferable when health, safety, and aesthetics issues are raised to challenge personal communication systems, since its low transmission powers, short link distances, low profile transceivers and limited spatial coverage make OW one of the least obtrusive personal network technologies. Finally, OW neither suffers from nor directly creates RF interference and OW systems can operate side-by-side with RF systems in personal communication networks. The high-speed electronics that drives OW transmissions must be designed not to be a indirect source of RF interference.

Apart from the generic requirements mentioned above, OW has been quite successful in personal communication systems since it exhibits attributes that closely match key requirements in these systems and their applications [2]. From an application perspective, applications like contact information exchange or file transfer between mobile devices are required to have a very short interaction time. In addition, several applications, for instance secure payment, require increased security and immunity to interference. The short distances involved and the limited spatial coverage of OW, which results from the narrow beams that are utilized, provide inherent security, and interference rejection, while at the same time allow for faster completion of the device discovery process that affects the application interaction time. Moreover, the abundant bandwidth of OW transceivers, more than 1 Gb/s at the time of writing of this article, facilitates speedy file transmissions and as a net result the total interaction time is minimized.

The suitability of OW for personal communication systems has drawn the attention of two key players in personal communication systems: IEEE and the Infrared Data Association (IrDA). IEEE had included an OW option in its original 802.11 MAC and PHY standards [5]. The standards defined a wireless communication scheme using diffuse optical channels at 1 and 2 Mb/s. However, they were never updated and are now rendered obsolete. Recently, IEEE launched a separate Task-Group within the framework of 802.15 (IEEE 802.15.7), which is responsible for producing MAC and PHY standards for OW communications based on visible light [6]. IEEE activities will be further discussed in the next section that details indoor systems.

The IrDA aims at equipping personal communication devices (computers, cellular phones, PDAs, etc.) with low-cost and high-bandwidth OW equipment. To this end, IrDA has produced numerous standards for the definition and implementation of (a) an infrared physical layer, (b) a protocol stack that is well suited for short-range OW, and (c) several application scenarios. Even though IrDA had produced standards for multiple-access implementations over infrared links (Advanced Infrared--AIr) [7–9], with possible applications in personal and local area networks (PANs and LANs), it soon became evident that it will be quite challenging for OW to compete with RF in terms of area coverage and user mobility. As a result, IrDA has focused primarily on point-and-shoot usage scenarios, since OW is well suited for this [10]: point-and-shoot applications require ultra-fast connection establishment and data transfer; both are readily offered by OW. The rest of the section focuses on the IrDA protocol stack, compliant applications and a comprehensive description of related research efforts.

2.1. The IrDA protocol stack and related applications

The IrDA protocol stack is shown in Figure 3. The lower layer is the physical layer [11, 12] (IrPHY), which operates at a nominal wavelength of 880 nm, although in practice infrared transceivers are allowed to operate in a range that extends from 850 to 900 nm. LEDs and silicon PIN photodiodes are utilized to reduce the system cost to an absolute minimum at lower speeds, while lasers are a more costly alternative at higher speeds. The average optical powers transmitted by lasers and LEDs are limited by eye and skin safety standards, according to [13]. Transmission (line) speeds vary from a couple of kb/s to 1 Gb/s and IrDA has provided detailed physical layer specifications for the following communication speeds:

- Serial Infrared (SIR): SIR is 16550 UART compatible, providing link speeds at 2.4, 9.6, 19.2, 38.4, 57.6, and 115.2 kb/s. The bytes that are to be transmitted over the infrared link are first placed in a 16550 UART chip, which is used to modulate the driving current at the transmitter LED. The UART chip appends a start and a stop bit to the transmitted byte, inverts the bit pattern and also alters the duty cycle of the '1's in the transmitted 10-bit UART frame to 3/16.

- Medium Infrared (MIR): MIR speeds are 0.576 and 1.152 Mb/s. Line coding is almost identical to the SIR speeds, with the only exception being the duty cycle of the '1's in MIR speeds (1/4 instead of 3/16).

- Fast Infrared (FIR): FIR supports a single line speed of 4 Mb/s. Line coding in FIR speeds follows a pulse-position-modulation (PPM) scheme with two-bit symbols. Four chips are required to transmit all possible symbols, and the symbol to chip mapping is defined as {00,1000}, {01,0100}, {10,0010}, and {11,0001}.

- Very Fast Infrared (VFIR): VFIR supports 16 Mb/s only. Line coding in VFIR follows a run-length limited HHH(1,13) scheme, which guarantees at least one empty chip after each pulse and allows for a maximum of 13 empty chips between pulses. Under the HHH(1,13) coding scheme, data bit pairs are encoded into three-chip symbols using the state transition table detailed in [11].

- Gigabit Infrared (Giga-IR): Giga-IR supports 512 Mb/s and 1.024 Gb/s. Line coding in Giga-IR is based on the 8010B DC-balanced encoding scheme, in which a data byte is encoded on a 10-bit character, so as to facilitate clock recovery at the receiver and to meet the stringer timing requirements of Giga-IR links. The Giga-IR optical transmitter can be either a LED or LD: if a LED is used as the transmitter then the encoded characters are sent over the infrared link using a four-level ASK modulation format, otherwise if a LD is used as the transmitter then the modulation format is a two-level ASK [12].

IrDA has also provided recommendations on a number of infrared link parameters. In SIR to VFIR speeds the maximum distance between devices should not exceed 1 m (or 200 mm for low power transceivers), while the 3 dB half-width angle ranges between 15 and 30° at the transmitter and 0-15° at the receiver. In Giga-IR speeds a docking station is typically used and the maximum distance is limited to 60 mm, while the transceiver angles are also limited to 5°.

Moreover, the bit-error-rate that should be achieved for all infrared links is no more than 10^{-8} for SIR to VFIR speeds and no more than 10^{-10} for Giga-IR, comparing directly to the BERs achieved in modern optical fiber networks. Additional optical and electrical parameters including (a) maximum/minimum optical intensity at the transmitter and optical irradiance at the receiver, (b) pulse rise/fall time, (c) pulse jitter, (d) minimum/maximum pulse duration, and (e) maximum allowable transmission rate deviations may be found in [11, 12] for all speeds ranging from SIR to Giga-IR.

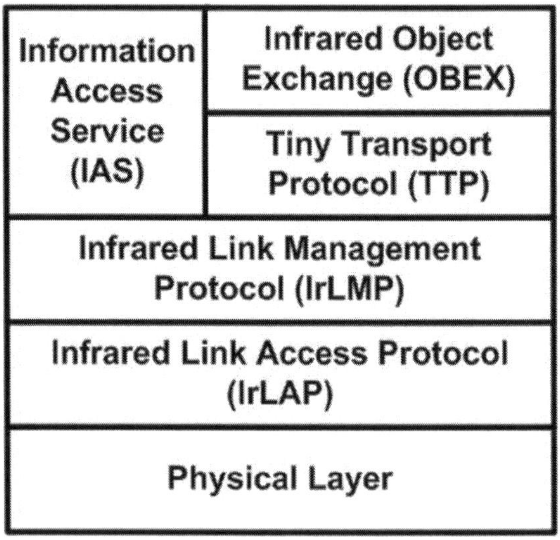

Figure 3 The IrDA protocol stack.

The IrDA protocol stack operates above IrPHY and consists of four main transfer protocols: the Infrared Link Access Protocol (IrLAP) [14], the Infrared Link Management Protocol (IrLMP) [15], the Tiny Transport Protocol (TTP) [16], and the Object Exchange (OBEX) protocol [17]. IrLAP is an HDLC-based link layer protocol with a primary function to provide reliable data transfer over half-duplex OW links. Earlier IrLAP implementations (SIR to VFIR speeds) operated in a Go-back-N fashion and maintained frame sequencing information in both directions. Currently, IrLAP is being modified to take full advantage of the bandwidth provided by Giga-IR speeds [18], and the next IrLAP standard is expected to provide (a) extended frame size, (b) extended window size, and (c) selective-repeat mode of operation. Apart from reliable data transfer at Gb/s speeds, IrLAP has additional features to account for the volatile nature of the infrared physical layer, including extended and dynamic node addressing, link recovery and rapid node discovery (25-300 ms) if fast-connect options have been implemented in IrLAP [19].

IrLMP operates above IrLAP and is responsible for providing upper layers with a multiplexing scheme over IrLAP, as well as a link control mechanism and a service discovery mechanism. IrLMP in fact comprises two separate entities, the Link Management Multiplexer (LM-MUX), which is responsible for link control and data multiplexing and (b) the Link Management Information Access Service (IAS) that maintains information about the services that have been registered at the local station. IrLMP communicates with upper layers and the IAS through multiple Logical Service Access Points (LSAPs). Each LSAP is assigned with a unique identifier that is called the Logical Service Access Point Selector (LSAP-Sel), and source/destination LSAP-Sel pairs are used to multiplex/demultiplex traffic from multiple IrLMP connections in LM-MUX. Moreover, LSAP-Sels of local services are registered in IAS to facilitate service discovery. Remote devices access the IAS by means of a simple command-response protocol (Information Access Protocol--IAP); the LSAP-Sel of IAS itself is well known and equals 0×00.

TTP is heavily associated with IrLMP and its main functions are (a) to provide per IrLMP connection flow control, and (b) segmentation and re-assembly (SAR) features. Even though IrLAP provides lower level flow control (back-pressure), it is not capable of distinguishing between individual IrLMP flows and as a result using IrLAP back-pressure will slow down all IrLMP connections. TTP on the other hand implements per-connection flow control by associating its own Service Access Points (TTPSAPs) with the LSAPs that are present in the LM-MUX on a one-to-one basis. The flow control mechanism itself is credit based: the TTP endpoints issue credits (up to 127 initially) to each other and they use up a credit whenever they send a TTP frame or receive a new credit whenever a TTP frame has been received at the other end. If a TTP endpoint reaches zero credits, it must wait for additional credits to be issued from the other side before sending additional frames.

OBEX is the session protocol that has been accepted by IrDA to ensure interoperability between devices, and provide a communication API for user applications. OBEX comprises three main components: (a) a model for representing objects, as well as information about objects, (b) a session protocol that enables session layer communication between devices, and (c) an application framework. Similar to HTTP, OBEX uses a header-based model to describe both the object attributes and the actual data of the object itself. Session peers, one acting as a client and the other as a server, exchange header information using a request/response communication protocol. Typically, OBEX operates in a stop-and-wait fashion, which means that only a single OBEX frame may be outstanding. This behavior limits the infrared link utilization, since requests and responses that involve the transmission of multiple OBEX frames need to wait for an equal number of link turnarounds. In order to mitigate this effect, enhanced OBEX implementations allow for a Single Response Mode (SRM), in which multi-frame requests/responses may be acknowledged by a single frame. Finally, the primary element of the application framework is the default OBEX server. This is the "well-known OBEX server" that is registered in the IAS database and provides basic functionalities like accessing file systems, databases, services and applications. The default OBEX server uses an

Inbox service, so that clients can exchange objects using a standard location. Alternatively, clients may also communicate with non-default OBEX servers; in such case multiple OBEX server instances are running along with the applications. The non-default server instances are registered with different LSAP-Sels in the IAS, so that there is no confusion with the default OBEX server.

IrDA has also standardized a stripped-down version of the protocol stack to facilitate implementations in limited capability terminals. This version is commonly known as IrDA Lite and maintains only the most fundamental operations of the protocols involved [20]. A more recent variation of the protocol stack, IrSimple [21], achieves faster transfer speeds by allowing blocks of information to be exchanged over IrLMP. Since the TTP flow control mechanism does not allow for block sizes over 127 frames, TTP is replaced by IrSMP (Sequence Management Protocol for IrSimple) in the IrSimple protocol stack [22]. IrSimple maintains the SAR capabilities of TTP and implements a frame sequencing mechanism to ensure the successful transmission of frame blocks.

IrDA has also provided standards for application related frameworks. One such application framework is Infrared Financial Messaging (IrFM), which supports digital payment applications running on infrared-enabled personal communication devices [23]. In IrFM, a personal communication device (also called a Personal Trusted Device--PTD) interacts with a Point of Sale (POS) terminal. Applications running on PTD and POS communicate over OBEX, with PTD being the OBEX server and POS being the OBEX client. Applications may not access OBEX directly, however, but only through IrFM services. To this end, the IrFM standard defines a number of core services that enable basic transactions between the POS and the PTD, as well as a Core IrFM Protocol (CIP) that passes transaction commands from IrFM services to OBEX. A key feature of IrFM is its fast transaction mechanism (point-and-pay), as a direct result of the point-and-shoot nature of the IrDA protocol stack. In addition, IrFM has been developed to maintain simplicity, security, interoperability and reliability in infrared financial transactions.

Another application framework is "wire replacement" and refers to the replacement of parallel and serial cables with IrDA links. So far there have been two relevant standards, the Serial and Parallel Port Emulation over IR (IrCOMM) [24] and the Infrared Universal Bus (Ir-USB) [25]. IrCOMM defines the emulation of parallel and serial ports over the IrDA protocol stack. The drive between the emulation is to facilitate the migration of applications that have been initially written to work over COM and LPT ports to IrDA compliant devices. To this end, IrCOMM differentiates between four service types: three-wire-raw, three-wire, nine-wire and Centronics. Three-wire-raw involves no control information and the role of IrCOMM is to fill up payload of IrLMP frames with user data. The other three service types, however, require a separate control channel and as a result IrCOMM frames carry control information along with the user data. In addition, IrCOMM interprets flow control commands and utilizes the credit based flow control mechanism of TinyTP in an appropriate fashion. Ir-USB, on the other hand, is mainly based on the USB 2.0 protocol stack and utilizes only IrLAP and IrPHY (Giga-IR in particular) as an ultrafast transmission mechanism. As a result, an operational speed of 480 Mb/s is provided to USB clients over a 512 Mb/s or 1.024 Gb/s infrared link. The critical subsystem of the Ir-USB architecture is the Ir-USB controller that acts as an interface between the electronic USB controller and IrLAP. Apart from transmitting data at very high speeds, the Ir-USB controller is also responsible for handling special purpose USB packets, so that the infrared link is essentially transparent to the electronic USB controller.

2.2. Research-related literature

OW personal communication systems involve point-to-point links with distances that are typically limited to a few tens of millimeters and very narrow optical beams. As a result multipath dispersion and fading, which severely affect the performance of indoor and outdoor systems, respectively, are not present in personal communication systems. The dominant impairment in personal communication systems is optical noise and interference from the sun and artificial lighting [26, 27], still the narrow reception angles filter out most of this. The reader may refer to Section III.D for a more detailed discussion on noise and interference, as well applicable mitigation techniques.

The optical channel in personal communication systems is considered to be linear, time invariant and non-dispersive [1, 28]. In practice, there are two main goals regarding the physical layer in personal communications systems: (a) to attain increasingly high link speeds, and (b) to deploy suitable codes that facilitate receiver functions such as clock and frame recovery. To these ends, simple amplitude modulation formats like OOK and amplitude shift keying (ASK) have been previously presented and evaluated [27], even though more complex phase and frequency modulation formats are also possible [29]. Relevant modulation schemes include PPM [30], differential PPM (DPPM) [31], digital pulse interval modulation (DPIM) [32], run-length limited (RLL) coding [33] and DC-balanced coding [34]. PPM is advantageous over OOK for low rate systems, since it requires decreased average power and is more robust against optical noise, especially noise components near baseband resulting from fluorescent lamps [27]. Still, PPM requires a bandwidth increased by a factor of $M/\log_2 M$ as compared to OOK, where M is the number of chips that are required to construct the PPM symbol. DPIM and DPPM require less bandwidth (approximately half) than PPM and more average power [30], but they also occupy more bandwidth than OOK and ASK. ASK is better suited for higher rate systems; it utilizes the same bandwidth as OOK to provide a $\log_2 M$ times faster line rate, at the expense of increased average optical power. Table 2 and Figure 4 summarize the bandwidth and BER performance of PPM variants and ASK systems in short-range OW.

Table 2 Bandwidth and BER performance of OOK and PPM variants

Modulation format	Data rate	BER
ASK	$R_b = B \cdot \log_2 M$	$2 \cdot \frac{M-1}{M \cdot \log_2 M} \cdot Q\left(\frac{1}{M-1} \cdot \sqrt{SNR \cdot \log_2 M}\right)$
PPM	$R_b = B \cdot \frac{\log_2 M}{M}$	$\frac{M}{2} \cdot Q\left(\sqrt{SNR \cdot \frac{M \cdot \log_2 M}{2}}\right)$
DPPM	$\overline{R_b} = 2 \cdot B \cdot \frac{\log_2 M}{M+1}$	Variable symbol duration, refer to [31] for a detailed presentation
DPIM	$\overline{R_b} = 2 \cdot B \cdot \frac{\log_2 M}{M+3}$	Variable symbol duration, refer to [32] for a detailed presentation

Bandwidth B is the first-null bandwidth, SNR is the electrical signal-to-noise ratio and M is the number of chips that construct a symbol in PPM or the number of power levels in an ASK system. Function $Q(x)$ is defined as $Q(x) = \frac{1}{\sqrt{2\pi}} \cdot \int_x^{\infty} e^{-u^2/2} \cdot du$.

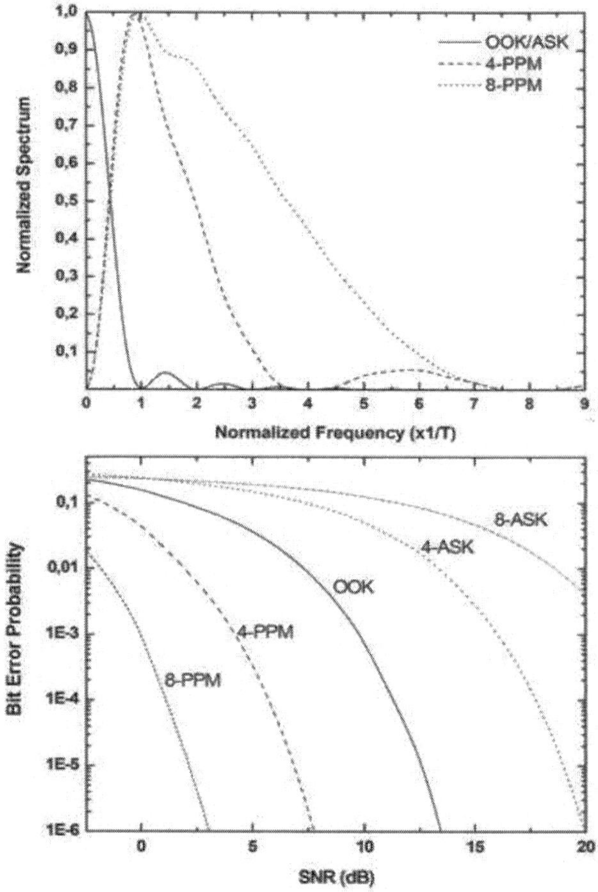

Figure 4 Spectral properties and BER performance of PPM and OOK/ASK systems, following the analysis of **[60]**.

RLL codes, on the other hand, are used to alleviate timing uncertainties that cause erroneous data recovery in high-speed PPM receivers. Specifically, the

VFIR IrDA HHH(1,13) run-length limited code performs better clock and data recovery than PPM, while maintaining an acceptable bandwidth efficiency (approximately 67%) [33]. DC-balanced coding is also receiver oriented and is used to mitigate the baseline wander effect caused by high-pass filtering of optical noise at the receiver [34–36]. More sophisticated modulation schemes that involve transmission over multiple carriers, such as subcarrier modulation (SCM) [37, 38] and orthogonal frequency division multiplexing (OFDM) [39–41] are also possible. However, SCM and OFDM are primarily designed to minimize inter-symbol (ISI) interference in indoor systems, and therefore require more complex coding and modulation hardware at the transceivers when compared to OOK.

With respect to protocol analysis in OW personal communication systems, both the initial IEEE 802.11 and IrDA AIr standards have been previously studied in the literature. A performance analysis for AIr may be found in [42–44] where the link-layer frame size is contrasted against the average frame delay and system throughput. IEEE 802.11 has been extensively investigated in the literature, but a complete overview is beyond the scope of this article. Two articles that evaluate the performance of initial 802.11, and apply to OW MAC and PHY are [45, 46]. As far as the point-and-shoot oriented IrDA protocol stack is concerned, numerous research efforts have been presented in the literature. A primary research topic involves the determination of the optimum values that maximize the system throughput for network parameters including (a) the OBEX frame size, (b) the TTP window size, and (c) the IrLAP window size, frame size and link turnaround time. The optimum IrLAP frame and window sizes have been calculated in [47, 48] as a function of the link BER, speed and turnaround time. Cross-layer optimization that involves the TTP, however, shows that the IrLAP network parameters are not independent of the TTP flow-control mechanism [49], and this owes to the fact that TTP frame streams are not temporally continuous. Moreover, the Stop-and-Wait fashion in which OBEX operates further complicates the determination of optimum values for the OBEX, TTP and IrLAP network parameters [50, 51]. Further analysis of the IrDA protocol stack involves several variations: the suitability and scalability of IrSimple for transmission of relatively small files over high-speed OW links is discussed in [52]. An alternative session protocol that is suitable for the transmission of larger files, IrBurst [53], is also optimized against the TTP and IrLAP network parameters in [54, 55]. Finally, a detailed analysis of IrFM attributes, including its connection delay (transaction speed), simplicity, interoperability, security and reliability, along with a comparison between IrFM and competing technologies may be found in [56].

3. INDOOR OW SYSTEMS

OW has been extensively studied as a broadband, low cost and power efficient technology for indoor communication systems [1, 27, 57–62]. Similar to the case of personal communications, OW indoor systems provide considerable cost savings to device manufacturers and end users due to the existence of unlicensed available spectrum, are perceived as safer than competing RF technologies to humans and do not suffer from RF interference. Indoor OW link architectures, however, are quite different from their personal communication counterparts due to the fact that a key requirement here is user mobility. In real-world indoor systems the user is allowed to move within a limited area of coverage, which typically equals the size of a room or an aircraft cabin.

Providing OW access to a moving user can be challenging, since the optical beams are blocked by objects inside the room. A possible solution is to deploy a roof-mounted and narrow-beam transmitter that is controlled by a tracking mechanism. The tracking mechanism rotates the transmitter and directs the narrow optical beam from the transmitter to the mobile receiver. This approach leads to a point-to-point OW link whose architecture is very similar to that of personal communication systems (Figure 5a). An alternative approach is to design the transmitter with a broad-angle emission pattern and allow multiple reflections from the walls and objects inside the room. In such a case, the OW system is categorized as diffuse (Figure 5b). If the emission pattern is properly engineered, the multiple reflections result in uniform illumination inside the area boundaries, even at places that are out of line-of-sight of the transmitter, and therefore enhance user mobility [63]. However, the multipath propagation in diffuse channels leads to time-dispersion of any pulse transmitted on the channel. Multipath dispersion is a critical impairment, since it drastically reduces the system bandwidth, and attainable data rate, and increases the link losses. As a result, point-to-point and diffuse topologies present a tradeoff between mobility and bandwidth. More advanced topologies, for instance multispot diffusing (Figure 5c), combine the best of both worlds and allow for significant user mobility while at the same time achieve two orders of magnitude higher bandwidth than purely diffuse ones. All three topologies (point-to-point, diffuse, and multispot diffusing) are discussed in further detail in Sections 3.1-3.3, and a summary of their respective strengths and weaknesses is given in Table 3.

Another major difference between personal communication and indoor systems is the impact of noise on their performance. Indoor OW receivers are typically designed with broader reception angles and as a result the received optical noise power levels are higher when compared to those of personal communications. Noise originating from the sun and artificial lighting interferes with the optical signal and degrades the signal quality at a baseband frequency range between DC and tens of MHz. This is unfortunately the operational frequency range for most indoor OW systems. Filtering techniques combined with modulation formats are required to reject a significant percentage of the received noise power and to limit the noise induced performance degradation. Noise sources, their impact on indoor OW and mitigation techniques are discussed in Section 3.4.

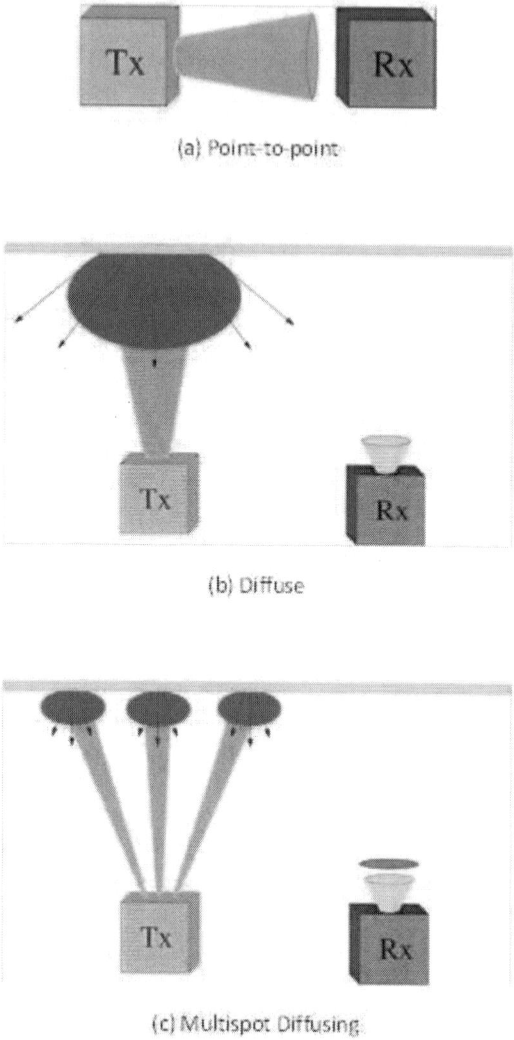

Figure 5 Topologies for indoor optical wireless systems. **(a)** Point-to-point, **(b)** diffuse, **(c)** multispot diffusing.

Finally, apart from discussing system requirements and key impairments in indoor OW systems, this section also presents the latest trend in indoor OW: VLCs. The main idea behind VLCs is to use solid state LEDs, which were originally developed for indoor lighting, for communication purposes, as well. VLCs are particularly appealing since they do not require additional transmitters and receivers, as is the case in infrared communications, and since ceiling LED installations are already designed to provide a uniform illumination pattern of high optical power within the area of coverage [64]. VLCs are presented in Section 3.5.

Table 3 Comparison of OW link topologies for indoor systems

	Link type		
	Point-to-point	**Diffuse**	**Multispot diffusing**
Link rate	High	Moderate	Moderate
Implementation complexity	Low	Low	High
Beam pointing	Required (exact)	Not-required	Required (partial)
Beam blocking	Yes	No	No
User mobility	Limited	High	High
Dispersion	None	High	Low
Path loss	Low	High	Moderate
Impact of ambient noise	Low	High	Low

3.1. Point-to-point links

Point-to-point links require a line-of-sight (LOS) between the transmitter and the receiver, as shown in Figure 5a. Such a link topology permits the use of highly directional transmitters with low beam divergence and narrow field-of-view (FOV) receivers. Thus, point-to-point links are able to reject the majority of ambient light that interferes with communications. In addition, the directionality of the link removes the impact of multipath dispersion since any reflected transmitter energy is essentially rejected by the receiver. Still, a disadvantage of this link topology is the fact that precise alignment and pointing are required, making the realization of mobile links difficult.

Very high data rates can be achieved in point-to-point indoor OW links by means of spatial division multiplexing, a technique that has been considered for some time. In [57], hundreds of parallel links, each operating at 50 Mb/s, were shown feasible in large areas using a ceiling-mounted transmitter. The transmitter was capable of creating a large number of narrow beams pointed to a variety of locations. Although such systems are ideal for high-speed data distribution, they are of high complexity and do not permit motion of terminals. High-speed electronic tracking was reported in [65] to support mobility at a rate of 155 Mb/s over ranges of nearly 2 m. Tracking was accomplished by using a transmitter that had an array of laser diodes combined with optics and a receiver with a wide FOV and detection array. Tracking involved the selection and switching of paths onto the appropriate receive element.

Point-to-point OW links have also found a great deal of popularity in a variety of shorter-range applications. As an example, the IrDA IrFM specification permits secure and quick payment options at point-of-sale terminals [56], as has been detailed in Section 2. Another example is a system provided by Talking Signs, which is able to deliver voice to visually impaired individuals using fixed infrared transmitters and hand-held receivers. Yet another area of application that has been active for some time is the use of OW links for high-speed backplanes. High bandwidth communication between circuit boards is a major limitation in many large computing and communication platforms. Using an array of tightly aligned transmitters and receivers, high-data rate links that are free from the capacitive, EMI and bandwidth limitations of on-board links [66, 67] can be established. An OW backplane has also been proposed to provide intra-craft data connectivity in a spacecraft [68].

The majority of indoor point-to-point OW links have single element transmitters and receivers. However, great gains can be attained by using multiple-transmitters and receivers in a coordinated fashion. The implementation of such Multiple-Input-Multiple-Output (MIMO) links involves the use of laser and photodiode arrays [69], optical interconnects [70] or a spatial-light modulator (SLM) as a transmitter and an imager as a receiver. Incoming data is either parallelized and transmitted over multiple beams, while when a SLM is employed, data is sent by creating a series of two-dimensional (2D) optical intensity images. OW MIMO links require tight spatial alignment between the transmitter and receiver arrays to avoid interference between the channels [71–73]. Recently a spatial processing technique that does not require tight spatial alignment has been developed for OW MIMO links [74, 75]. In this technique, the interference between spatial channels is modeled and used in communication system design directly to provide high spectral efficiency. Alternatively, interference can be minimized by allowing only a single laser element to transmit, as proposed in [76], or by using interference cancelation techniques at the receiver [77].

MIMO links have been used in a variety of applications including 2D barcodes [78, 79], page-oriented recording [80, 81], holographic storage [71, 72, 82, 83], and MIMO OW communications [27, 66, 84, 85]. With respect to the 2D barcodes, OW MIMO links are currently used on a wide array of cellular telephones in the form of barcode readers. Data are encoded as a 2D barcode which is either printed or displayed on a billboard. A cell phone camera is then used to image the barcode and processing is done to extract the data. Many newspapers now publish 2D barcodes which, when imaged by a cell phone camera, direct the cell phone browser to the appropriate webpage for additional details on the story [86].

3.2. Diffuse indoor OW links

Initial studies into diffuse OW links were presented in 1979 [1] and diffuse links were shown to be robust, immune to shadowing and allowed for terminal mobility. Unlike point-to-point OW links, a diffuse OW link does not require the existence of a LOS to establish communications. Rather, as illustrated in Figure 5b, it employs wide beam transmitters and large FOV receivers to exploit diffuse reflections from the walls and objects inside the room and establish a communication link [1, 27, 87, 88]. As a result, there is inherent path diversity and diffuse links are relatively immune to blockage and pointing errors and permit a great degree of mobility for the receivers. However, the received signal is corrupted by multipath dispersion due to the large number of collected reflections at the receiver. The temporal spread of reflections results in inter-symbol interference at high data rates and limits the system bandwidth. Notice that although diffuse OW links suffer from multipath dispersion, they do not exhibit multipath fading as in RF links. The detector in diffuse OW links typically has a size of many thousands of wavelengths and the received optical intensity is integrated over the surface of the detector and provides an inherent degree of spatial diversity [27]. Thus, the channel characteristics are unaltered when moving the detector by many wavelengths.

The indoor diffuse OW channel is well modeled as a linear, time-dispersive channel with a time-invariant impulse response $h(t)$ [89]. The impulse response is calculated from the individual reflections of walls and objects inside the room and there is an array of methods in the literature to compute the impulse response in optical intensity domain for a typical room layout. These methods include ray tracing [89, 90], experimental measurements [28, 91, 92], simplified functional modeling [93, 94], photon tracing [95] and Monte Carlo ray tracing methods [96–99]. While specular reflections are possible, the vast majority of reflections are diffuse since the dimensions of the surface roughness is on the order of the wavelength of light used in the link [95]. As a result, the optical power is reflected nearly independently of the angle of incidence of the optical wave. The intensity distribution of an individual diffuse reflection is well approximated by a Lambertian pattern [1, 60, 89].

The impact of multipath dispersion on diffuse indoor OW links is quantified by the path loss and normalized delay spread. The path loss is calculated from the channel impulse response $h(t)$ and ranges from 10^{-7} to 10^{-5}[27, 28]. The temporal dispersion due to multipath can be measured by the root-mean square (RMS) delay spread of the impulse response and ranges from 2 to 10 ns [27, 28]. A popular and convenient model that correlates the temporal dispersion with the channel bandwidth is the exponential-decay model [93, 94]. In this model, the 3-dB bandwidth of the channel impulse response is calculated from the channel temporal dispersion D using the following simple analytical relation

$$f_{3\text{dB}} = \frac{1}{4\pi D}. \tag{3.1}$$

A range in temporal dispersion D from 2 to 10 ns [27, 28] implies that the 3-dB channel bandwidth approximately ranges from 10 to 40 MHz. The impulse response, delay spread and path loss depend on the particular composition and arrangement of the room and are considered fixed once the room characteristics are fixed [27].

Multipath dispersion severely limits the diffuse channel bandwidth and a vast array of modulation, coding and equalization techniques have been applied to combat this effect [27, 89, 100–103]. An important aspect of these techniques is that amplitude constraints must also be imposed on all signals transmitted on OW channels, since the optical power signal must satisfy a non-negativity constraint. In addition, due to eye- and skin-safety requirements a limitation on the average optical power is also mandatory as imposed by international eye-safety limitations [13]. This power limitation is in stark contrast to the mean squared constraint in RF channels. As a result, the use of modulation and coding for RF channels directly on OW channels is seldom efficient and the unique power and bandwidth efficiency requirement of the OW channel must be considered in any communication system design [60].

Diffuse OW links have been standardized by both IrDA and IEEE. The standardization of diffuse OW links from IrDA resulted in the Advanced Infrared (AIr) physical layer interface that defines the operation of 4 Mb/s diffuse OW links inside typical room sizes [2, 43, 44, 104]. IEEE included a physical layer specification for infrared OW communications at 1 and 2 Mb/s in its 802.11 standard. Experimentally, diffuse infrared OW links have been demonstrated at 25 Mb/s [105] and 50 Mb/s [106, 107].

3.3. Multispot diffusing links

Spatial diversity techniques have been successfully applied to indoor OW links in a topology that is termed multispot diffusing (MSD) and shown in Figure 5c[108]. The MSD architecture aims to combine the advantages of point-to-point links with the mobility of diffuse links [108–115]. To this end, the MSD transmitter modulates data onto a series of beams that are projected onto the ceiling above the communications floor. The MSD receiver ideally images one or perhaps several spots and decodes data from the diffusely reflected energy. The same data is modulated for all spots and the arrangement and number of spots is optimized so that at least one spot is in the imager for every receiver position.

Even though both diffuse OW and MSD links rely on diffuse reflections on surfaces inside the room, there are large differences in the underlying channel models of both topologies. The first main difference lies in the intensity distribution of the optical transmitter. Whereas diffuse OW links emit light over a large divergence angle, the MSD transmitter uses a series of narrow divergence beams directed to the ceiling. This more efficient organization of the optical intensity leads to a far smaller path loss in MSD systems than in diffuse systems. The second major difference lies in the construction of the receiver. The MSD receiver consists of a series of narrow FOV elements directed to the ceiling. These narrow FOV receivers reject a large portion of ambient light. In addition, the narrow FOV receivers are able to reject any stray multipath signals and this effect significantly improves the bandwidth of the channel. It has been shown that in a typical room setting the channel bandwidth of an MSD system is in excess of 2 GHz, which is in stark contrast to the tens of MHz available in diffuse systems [116].

One of the challenges in implementing an MSD link is the creation of an array of spots on the ceiling. Although straightforward, using arrays of emitters with their own optics is bulky and expensive [115]. A less expensive and compact method is to use a laser diode and a computer generated hologram. In [112], a hologram which produces an array of 8 × 8 beams is presented, while holographic diffusers with fewer lobes were considered earlier in [63]. Holograms have also been considered to create arrays as large as 10 × 10 spots [117]. Other work has considered different geometries for MSD spots including a line strip and a diamond in an effort to reduce transmitter complexity [118, 119]. Experimental work on MSD systems has also shown promising results. An angle diversity receiver composed of discrete elements working at 70 Mb/s at a BER less than 10^{-9} over a 4-m range was reported in [115]. In [114], system issues for a 100 Mb/s MSD link were also presented. A 140 Mb/s MSD system with a single receiver [120] and a 155 Mb/s using a multi-element imaging receiver are presented in [65]. Following the above, MSD links maintain the high bandwidth and low path losses of point-to-point links and at the same time achieve mobility that is comparable with those in diffuse links at the expense of slightly more complicated transmitters and receivers [112]. As a result, the MSD architecture is an excellent topology for future indoor high-speed content distribution systems.

3.4. Noise and interference modeling and mitigation

Noise and interference are two additional impairments that affect the performance of indoor OW links. In typical indoor diffuse OW systems, which involve intensity modulation and direct detection (IM/DD), a good model that describes noise and interference is the baseband model

$$y(t) = R \cdot x(t) * h(t) + z(t)\,. \tag{3.2}$$

$x(t)$ is the transmitted optical power, $y(t)$ is the current at the receiver, $h(t)$ is the optical channel impulse response, R is the receiver photodiode responsivity and * denotes continuous time convolution. The signal $z(t)$ represents noise and interference added at the receiver. In general, $z(t)$ is a high intensity shot noise process. Such processes are well modeled as zero-mean, Gaussian distributed, independent of $x(t)$ and white [27]. However, this model is only true under the assumption of intense ambient light. For example, in links with low background illumination a photon-counting model is more appropriate [60].

The sources of noise and interference include background illumination, lighting from incandescent and fluorescent sources as well are front-end circuit noise in the preamp of the receiver [1, 26, 27, 60, 87, 121, 122]. Figure 6 summarizes the spectral distribution of solar, incandescent, and fluorescent radiation in the visible and near-infrared wavelength ranges, which are typically employed in OW systems. The sun emits broadband un-modulated light that is converted to shot-noise current at the receiver. The root-mean-square (RMS) of the noise current is proportional to the incident optical power and the receiver bandwidth and band-limited optical and electrical filters at the receiver can reject a significant portion of the shot noise. Incandescent lamps are modulated by their AC power supply (50/60 Hz) and produce a narrowband photocurrent with a total bandwidth of less than 2 kHz [122]. In practical OW systems the receiver bandwidth is significantly higher than 2 kHz and the photocurrent from incandescent light is modeled as a constant RMS shot noise current that is added to the photocurrent from sunlight [27]. Fluorescent lamps, on the other hand, emit light that is modulated at the power supply frequency (50/60 Hz) or higher frequencies (tens or hundreds of kHz). Light from 50/60 Hz fluorescent lamps produces a photocurrent with a total bandwidth of 20 kHz [122], while higher modulation frequencies result in photocurrents that have a considerable part of their power spectrum in the MHz range [121]. Consequently, the bandwidth of the photocurrent from fluorescent lamps is a significant percentage of the total bandwidth in practical OW systems and the photocurrent is modeled as a time-varying interference signal [27].

The effects of fluorescent light are particularly deleterious for OOK/ASK modulation formats, in which the electrical power spectrum of OOK/ASK contains a strong DC component. Fluorescent noise cannot be filtered without rejecting a significant portion of the signal in low rate OOK/ASK systems [121], and adaptive filtering [123–125] or spread spectrum [126] techniques are required. Higher rate OOK/ASK systems have a broader spectrum and fluorescent noise may be rejected through high pass filtering, however this leads to a baseline wander effect and additional DC-balanced line coding is required [35]. Low rate PPM systems and their variants (DPIM, DPPM) are less vulnerable to noise from slowly modulated fluorescent lamps than OOK/ASK systems, since the peak spectrums of PPM is located at non-zero frequencies and high-pass filtering rejects fluorescent noise without imposing significant ISI [121]. Rejecting fluorescent light modulated at higher frequencies, however, requires processing techniques similar to those used for OOK/ASK [123]. Finally, SCM and OFDM systems may avoid fluorescent noise altogether by utilizing transmission frequencies that are located outside the fluorescent noise bandwidth.

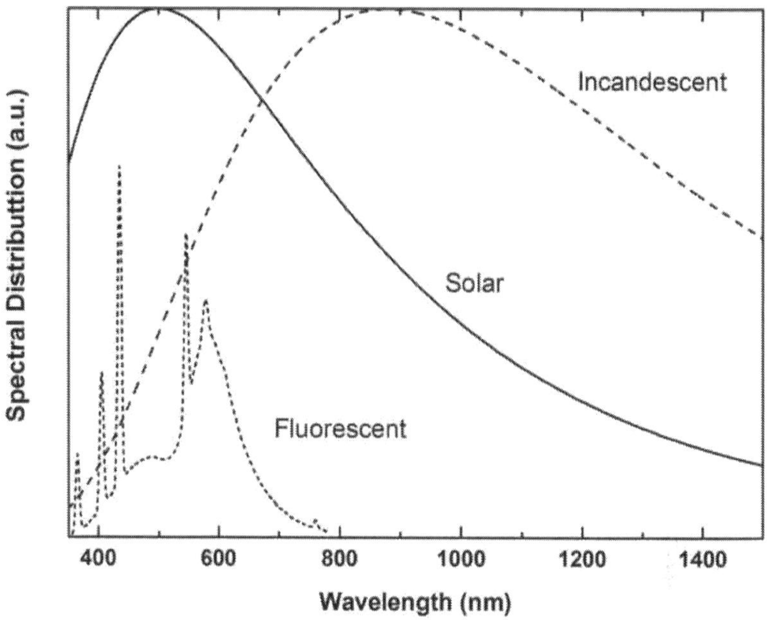

Figure 6 Spectral distribution of solar radiation and emissions from incandescent/fluorescent lamps in the visible and near-infrared wavelength range. Solar and incandescent spectra are calculated from Wien's displacement law at temperatures of 5800 and 3300°K, respectively. Fluorescent spectrum corresponds to a commercially available lamp.

3.5. VLC systems

With the advent of solid-state lighting based on white LEDs, there is intense interest in the use of visible light for OW communications. Not only do white LEDs for illumination purposes have high energy efficiency and lifetimes, but they also possess significant modulation bandwidth. By modulating the illumination outside of the human perceptual range, significant data rates can be transmitted while satisfying the illumination requirements inside a room [59, 127]. Such VLC channels also possess an SNR that is approximately 30 dB higher than a comparable infrared link since the background interference is largely removed [128]. Moreover, it is not necessary to install additional LEDs for telecommunication purposes. Visible light LEDs are already designed to provide uniform indoor lighting, which means that the effects of shadowing are minimized. In their simplest implementation VLCs only provide broadcast download capability, since the LED lighting will not in general include optical receivers.

VLCs have drawn significant attention from standardization bodies and research institutes in Europe, the USA and Japan. In Japan, VLC standardization efforts have been underway for some time. The VLC consortium [3] was founded in 2003 to investigate and provide a standard platform for VLC systems. In 2007, the VLC standard JEITA CP-1221 was issued by the Japan Electronics and Information Technology Industries Association [129]. European efforts in VLC have also been significant. The OMEGA project [130], funded by the European Commission [131], is tasked to develop the physical channels to enable Gb/s rates for home area networks combining RF, VLC, infrared and power line communications. A total of 20 partners are included in this project which has a budget on the order of €19 million and a European Commission fund of €12.4 million [132]. In the USA, an NSF research fund of $18.5 million has been provided to three universities to establish Smart Lighting Engineering Research Centers, which will develop device and system technologies for VLC communications. In addition, standardization efforts for VLC through the IEEE have been quite active recently [133].

Similar to infrared OW, the transmission rates of VLC systems are limited by the effects of multipath dispersion and noise from the sun and artificial lighting. In principle, the same techniques that have been presented in previous paragraphs to mitigate dispersion and noise may be applied to VLCs. As a result, multiple carrier modulation formats (SCM [134] and OFDM [135, 136]), MIMO systems [137], coding schemes [138, 139], and adaptive equalization methods [140] have been proposed in the literature within the context of VLCs. The transmission rates of VLC systems are also limited by the modulation bandwidth of phosphor-based visible LEDs; the slow time constant of the phosphor they are manufactured with in order to emit yellow light leads to a typical modulation bandwidth of a few MHz. Filtering and equalization at either the transmitter [141] or the receiver [142] alleviates the slow phosphor response, and increases the visible LED bandwidth by an order of magnitude. This bandwidth increase, along with the use of discrete multi-tone OFDM, has enabled the demonstration of VLCs operating at transmission rates that exceed 100 Mb/s [143–145].

An additional aspect of VLCs is that visible LEDS are also required to provide lighting at a constant power level and brightness adjustment options. The human eye is not capable of distinguishing between power variations at the bit level for typical transmission speeds under consideration in VLC systems (tens to hundreds of Mb/s). However, long series of '1's and '0's lead to long-term power variations that are perceived as flickering. Flickering is harmful for the human eye and several approaches, which aim at maintaining a constant DC-component in the emitted VLC signals, have been taken to mitigate its effect. The most common mitigation technique is to use PPM, in which transmitted symbols maintain a pre-determined percentage of '1's and '0's. A performance analysis of PPM variants may be found in [134, 146]. Another approach involves using DC-balanced line coding such as modified 4b5b, 5b6b or 8b10b over OOK [147]. As far as brightness adjustment is concerned, it is accomplished by combining line coding with pulse-width-modulation (PWM) variants [147, 148]. Under this scheme, line coding provides a DC-component whose amplitude can be adjusted by varying the duty cycle of the PWM.

4. OUTDOOR OW SYSTEMS

Another major application area of OW is in outdoor systems. From a commercial perspective, outdoor OW complements optical fibers in LAN interconnections and broadband access networks, providing a high-bandwidth and low-cost alternative [149]. The installation of an outdoor OW system alleviates the most significant expenditure of fiber deployment: the digging up and repairing of roads. In addition, current commercially available systems operate at 10 Gb/s Ethernet speeds, which are comparable to the speeds attained per channel in commercial optical fiber networks.

Directional OW links also have a low probability of intercept and detection compared to broadcast or less-directional RF wireless links. These are desirable attributes for certain applications, especially military communications, where the communicating parties need to remain undetectable by an adversary. In contrast to RF transmissions, where the adversary needs only a low complexity receiver to detect broadcast transmissions or the side-lobes of directional RF antennas, implementation of an OW link in a completely covert fashion requires fully optical pointing, acquisition and tracking without any adjunct RF technology from the third party. Thus, it is essentially impossible for the third party to detect the existence of the link, unless they insert their own receiver into the beam path. The security provided by the optical link is enhanced by means of encryption, which ultimately prevents interception of the actual data being transmitted.

Two key requirements for outdoor OW systems are ever-increasing line speeds and link distances. Multi-Gb/s OW line speeds are feasible at modest distances at the time of writing, owing to technological advances in optical fiber technology and adaptive optics. Fiber networks are expected to advance to 100 Gb/s in the near future, providing the necessary end-point components for OW, as well. As a result, OW line speeds are expected to keep up with their fiber optic counterparts, attaining ultra-high capacities per channel.

In ultra-long range outdoor OW systems, however, the achievable data rates can be severely limited by the earth's atmosphere. The atmosphere has a detrimental impact on optical beams, an impact that worsens as the link length increases [150]. The effects that come into play in making the outdoor OW propagation channel non-ideal are shown schematically in Figure 7. Obscuration (haze, fog, snow, rain) along the path reduces the received power. Even in the absence of significant obscuration, atmospheric turbulence presents significant challenges in the operation of OW links for distances as short as 500 m. Turbulence causes fluctuating intensity at the received power and beam wandering, spreading or breakup. The net result of these effects is manifest as scintillation in the received optical power. Scintillation presents a challenge for the receiver in an OW link, since the receiver must have significant dynamic range, which may need to be more than 40 dB, to handle large variations in the received intensity. Reductions in the received power that take the receiver below the specified threshold correspond to fades. During a deep fade, receiver and background noise dominate over the received signal, rendering the link useless.

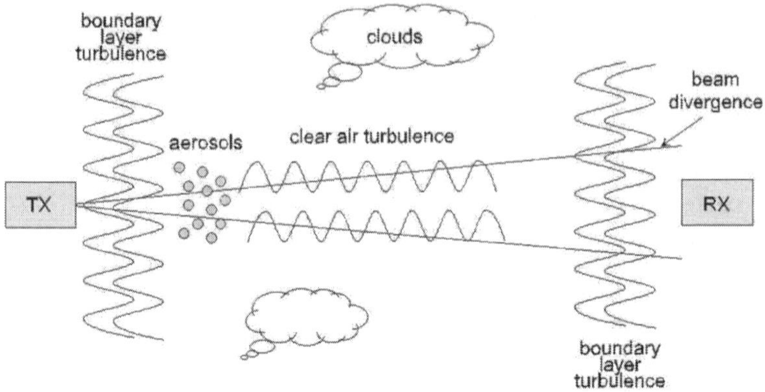

Figure 7 Factors affecting the communication channel for outdoor FSO links.

The stochastic nature of the fades and surges, both in terms of their intensity and their duration, leads to a stochastic description of the channel response. This is in stark contrast with personal communication and indoor systems, where the channel impulse response is modeled as time invariant. Stochastic channel models suitable for outdoor OW are detailed in Section 4.1. Moreover, due to the stochastic nature of the outdoor OW channel, mitigation of channel-imposed degradations is challenging. Applicable mitigation techniques involve some form of diversity, either spatial or temporal, to increase the probability of signal reception. This is accomplished by sending multiple replicas of the original signal over different propagation paths (spatial diversity) or by temporally displacing replicas by an offset that exceeds the fade duration (temporal diversity). Error-correction-coding is also applicable, but in high capacity systems the number of bits affected by a fade increases with the line rate. Still, these techniques can easily add several seconds of latency to a link. In the Gb/s regime, the code length that is required to correct the erroneous bits might be impractically long. Mitigation techniques are discussed in Section 4.2.

The section concludes with the most challenging area of application of outdoor OW systems, satellite communications. These systems employ laser beams that communicate data between terrestrial feeders, satellites and airborne relays and involve link distances that reach up to thousands of kilometers. The impact of atmospheric effects is magnified for such distances, while highly accurate beam acquisition, pointing and tracking systems are required to maintain a working link. OW for satellite communications is presented in Section 4.3.

4.1. Channel modeling

A major effect of the atmosphere on optical beams is the rapid attenuation of their power when they propagate through obscurations (haze, fog, snow and rain). For a single mode Gaussian beam of small angle beam divergence θ and power P_{TX} at the transmitter, the received power at an aperture of area A at range L equals

$$P_{RX} = \frac{2P_{TX}Ae^{-\alpha L}}{\pi\theta^2L^2}, \tag{4.1}$$

where α is the intensity attenuation coefficient along the obscured path. Typical obscuration losses that are encountered near the ground range from 0.5 dB/km in clear air, 3 dB/km in haze, 50 dB/km in thick fog, and 350 dB/km in impenetrable fog (OW links are more affected by snow and fog than they are by rain). At high altitude in clear air the attenuation may be as low as 0.1 dB/km, although aerosols can still affect the channel [151]. In practice, the exponential decay of the optical power means that OW links do not work over significant distances through obscuration and it is not practical to attempt to operate OW links through significant obscuration by increasing transmitter power. Some additional range can be accomplished by decreasing beam divergence angle θ, although this imposes more rigorous pointing and tracking requirements on the link.

A second performance degradation in outdoor OW systems results from the rapid scintillation of the received power caused by atmospheric turbulence. The local density of the atmosphere is constantly fluctuating because of temperature and pressure fluctuations. When a laser beam propagates through the atmosphere, the randomly varying spatial distribution of refractive index that it encounters causes the following effects [152]:

(a)Fluctuating intensity as observed with an optical detector at the end of the path (scintillation).

(b)A varying degree of intensity fluctuation that depends on the receiver aperture (the size of the detector, or the size of the receiving optics which direct the collected light to the detector).

(c)Progressive deterioration of circularly symmetric Gaussian beams with increasing distance and turbulence strength. The progressive changes that are observed are time-dependent deviations of the beam shape from circular, wander of the centroid of the beam, increase in the width of the beam over and above that expected from diffraction, and breakup of the beam into distinct patches of illumination whose shapes and locations fluctuate with time.

(d)A fall in the coherence length of the laser beam.

(e)Fluctuations in the angle of arrival of the phase fronts at the receiver.

Theoretical analysis of the effects of the turbulent atmosphere on the propagation of electromagnetic waves has been extensive, and to some extent repetitive, over at least the last 40 years. A substantial amount of this study was carried out in the former Soviet Union, and the books by Tatarski [153] and Chernov [154] are classic accounts of the state of the field at the beginning of the 1960s. Comprehensive reviews of the specific situation regarding laser beam propagation have been given by Ishimaru [155], Strohbehn [156], Zuev [157], Andrews and Phillips [152], and Andrews et al. [158]. There are also a number of other valuable review articles, for example [159, 160]. Many hundreds of additional articles covering the effect of turbulence on propagating laser beams can be found in [152].

The turbulence induced scintillation at the receiver is characterized by the probability density function (pdf) of the received power. The pdf allows the calculation of the probability of fades or surges of a given level below or above the average received power. The pdf can also be used to estimate the fraction of time that a given link will be in a fade or surge state. It is frequently found experimentally that over a wide range of scintillation indices the distribution of log field amplitude is a normal distribution, as are the log intensity variations measured at a point [156]. For log-normal statistics the probability of a fade (or surge) of a given depth can be calculated using the methods described by Yura and McKinley [161], who also provide a means to calculate the mean duration of a fade. An alternative pdf that better describes the distribution of received power in strong turbulence is the γ-γ distribution [157, 159, 162, 163]. If scintillation is reduced by aperture averaging, which is likely for all but very long links, then log-normal statistics are usually adequate to describe the behavior.

The level of scintillation is taken into account in individual pdfs by the scintillation index

$$\sigma_I^2 = \frac{\langle I^2\rangle - \langle I\rangle^2}{\langle I\rangle^2}. \tag{4.2}$$

I is the intensity measured by a point receiver and angle brackets indicate time averages. In practice, the receiver of an OW link will have a finite collection area and will only act as a point receiver if its diameter D is small enough to satisfy $D << \sqrt{\lambda L}$, where λ is the wavelength used and L is the length of the link. If the receiver diameter is larger than that of a point receiver then the level of scintillation is reduced by aperture averaging and the scintillation index in this case is given by Equation (4.2) with I replaced by the received power P.

The strength of turbulence is frequently characterized by the Rytov variance. The Rytov variance is obtained by approximating the solution to Maxwell's equations for electromagnetic wave propagation through a random medium under weak turbulence conditions. For a plane wave, the Rytov variance is calculated as

$$\sigma^2_{\ln I_R} = \left\langle \left(\ln I - \langle \ln I \rangle\right)^2 \right\rangle = 1.23 \cdot C_n^2 k^{7/6} L^{11/6}, \tag{4.3}$$

where C_n^2 is the refractive index structure constant and k is the wavevector magnitude of the light. In weak turbulence, when $\sigma 2I$<0.3, the Rytov approximation is valid and the scintillation index practically coincides with the Rytov variance. As the turbulence strength increases, however, the Rytov approximation no longer provides a precise solution to Maxwell's equations and the Rytov variance increases without limit. Strong turbulence theory is then required to predict the scintillation index behavior at increasing values of the Rytov variance and it has been shown that the scintillation index saturates at values of the Rytov variance that can range up to 5-6 [158]. Figure 8 show the behavior of the scintillation index, calculated from weak and strong turbulence theory [160], as a function of the Rytov variance.

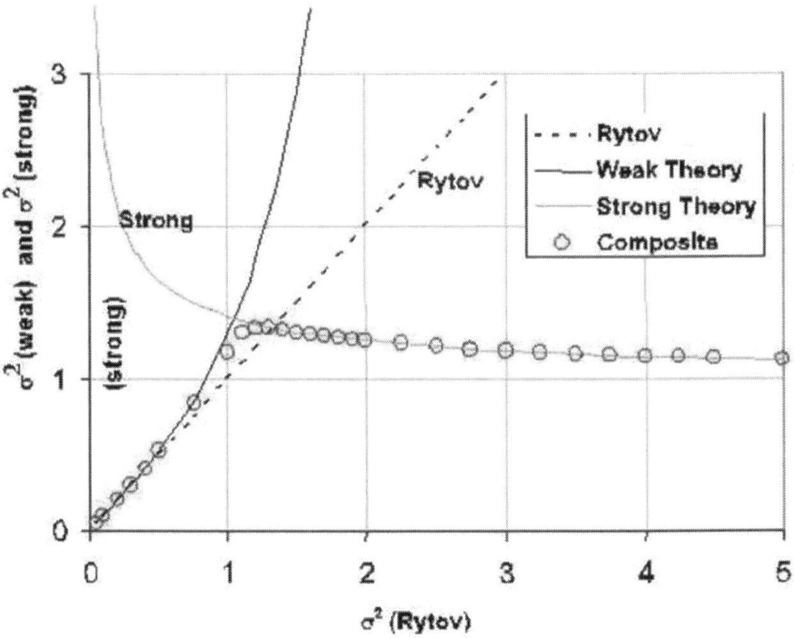

Figure 8 Comparison of log intensity variances calculated using weak and strong turbulence theories **[160]**for an atmospheric path of 863 m at 632.8 nm.

The aforementioned atmospheric impairments are combined with factors that include (a) the optical power losses in the transmitter and receiver, (b) beam spreading, (c) the Strehl (optical imaging quality) ratio, and (d) pointing errors to calculate link parameters such as the link margin and link availability [164–166]. The link margin is defined as the ratio of the transmitted power to the minimum power that is required to achieve a desired BER at the receiver. Signal fades that exceed the link margin lead to an increased BER for the duration of the fade and extremely deep fades raise the BER to 50%, causing a total loss of data and rendering the link useless. In contrast to other wireless channels, signal fading on the OW channel is relatively slow compared to the data-rate. The duration of a fade might be up to 100 ms or longer when the clock period, which depends on the data rate, is generally less than 20 ns (50 Mb/s). Hence, during a single fade several thousand transmitted bits can be corrupted. The link availability, on the other hand, is measured over a larger timescale and is defined as the percentage of time over a year that the link is available. The link availability is primarily dependent upon the existence of obscurations in the atmosphere. Under heavy obscuration, the atmospheric losses exceed the link margin and, similar to deep fades, the link operates at an increased BER or is unusable for the duration of the obscuration. When link outage resulting from obscuration cannot be tolerated the best solution is to operate a hybrid link where the OW link is operated in parallel with a directional RF link in unlicensed regions of C-band (5.8 GHz) or Ku-band. Hybrid OW/RF systems are detailed in the next section.

4.2. Fade mitigation techniques

(1) There are four principal methods for mitigating the effects of fading at the physical layer: aperture averaging, diversity, error correction, and adaptive optics.

Aperture averaging: Aperture averaging reduces the scintillation index by using a receiver whose aperture diameter D is larger than the lateral correlation distance on the link $\sqrt{\lambda L}$. Measurements of the effect of the aperture size on scintillation index have been given in [167]. These measurements, made in intermediate strength turbulence, are in reasonable agreement with the weak turbulence formula given in [152]

$$F = \frac{\sigma_I^2(D)}{\sigma_I^2(0)} = \left[1 + 1.07 \cdot \left(\frac{kD^2}{4L}\right)\right]^{-1}, \tag{4.4}$$

which describes the reduction F in the scintillation index for a receiver of diameter D compared to a point receiver and holds for plane waves with a small inner scale l_0 satisfying $l_0 << \sqrt{L/k}$. The aperture averaging factor does not depend on the exact shape of the aperture and depends primarily on receiver aperture area [168].

(2) Diversity techniques: Diversity techniques use multiple identical transmissions from sources to receivers [169, 170]. The multiple transmissions take uncorrelated paths and as a result the effective level of scintillation and the probability of a fade can be reduced. The most common diversity approach uses multiple transmitter and receiver apertures (MIMO), which are spaced further apart than the lateral correlation distance on the link [171–173]. If n independent apertures are used and if the received signals are logically combined then the probability of a fade of a given depth is reduced to prob $(P \leq P_{\text{min}})^n$, with P_{min} being the minimum power at each receiver aperture. For a fair comparison of the diversity approach the probability should be evaluated for the same total power transmitted, which means that the fade reduction is really $\text{frac}\left(\frac{P_{\text{Total}}}{n} \leq P_{\text{min}}\right)^n$, where frac($x$) denotes the fractional part of x.An alternative approach to spatial diversity is to use time-delayed diversity (TDD), described in [174, 175]. In this approach, n total transmissions of the same data take place between a single transmitter and a single receiver. The replicated transmissions are delayed by an offset τ which is longer than expected fade duration. The replicas are re-synchronized, buffered and logically combined at the receiver. If the separate transmissions of data are truly independent then the BER is reduced from its single channel value of BER to BER ^{n}. In practice, the exact reduction in BER must be calculated from the joint pdf for the signal received at different delay times, which requires knowledge of the temporal covariance matrix. Experimental measurements [175] have shown that in weak turbulence a delay of about 20 ms provides decorrelation between successive transmissions. The required decorrelation time becomes shorter as turbulence gets stronger, which is precisely when diversity approaches of all types become most valuable.

(3) Error correction: In some scenarios and especially for airborne terminals, size, power and weight constraints may not allow for multiple transmissions and therefore diversity cannot be implemented. Additionally, diversity systems can get very complex because of issues such as optics alignment, signal synchronization, interference effects, and others. Error correction is a simpler fade mitigation technique that has been shown to give satisfactory results [176] on SISO (single input single output) channels. Forward-error-correction (FEC), in particular, adds additional bit patterns to each coded packet being transmitted, which allows all transmitted bits to be recovered if some are lost due to fading. For example, in a (255,223) Reed-Solomon code that uses 8-bit symbols, there are 255 codeword bytes. Each codeword contains 223 bytes of data and 32 bytes for parity checking. The decoder can correct any 16 symbol errors in the code word: errors up to 16 bytes anywhere in the codeword can be automatically corrected. Still, this method only works well if fades are of short duration. In this example, the fade duration should be limited to 128-bit periods. Unfortunately, the transmission time of a code word in a high data rate OW system is much shorter than even the shortest fades. At 1.25 Gb/s, which is a commercially available OW data rate, 255 bytes takes only 1.6 μs for

transmission. As a result, improved coding and interleaving techniques are being investigated for application in higher rate OW systems.

Recent study has suggested that BER improvement can be obtained by the use of low density parity check (LDPC) codes [163]. It is not clear, however, that in this work the temporal characteristics of the strongly fading channel were properly accounted for. The performance of a link-layer LDPC staircase FEC system (these codes are called IRA - Irregular Repeat-Accumulate codes [177]) has also been tested over a Fast-Ethernet OW link [176]. In the presented test scenario the mentioned code was able to overcome all packet losses (11% of lost data). Another approach is to employ interleaving in parallel with coding. Interleaving provides a temporal diversity by spacing coded symbols by time intervals that are longer than expected fade durations. As a result, a fade only affects a single symbol per word in an interleaved in system and the length of the required code is drastically reduced. Further details about coding and interleaving for OW channels can be found in the specialized literature [177–180].

The drawback of FEC, and especially interleaving, is increased latency. The data have to be buffered at the transmitter and the receiver for at least the duration of the used coding constraint length. A rule of thumb suggests using block lengths that have a constraint length of several times the maximum expected fade duration. The processing time needed for encoding and decoding must also be added to the total processing delay. In general, the overall transmission delay can be expected to exceed 100 ms and therefore it is not practical to implement FEC-interleaving schemes for delay-critical data. Latency can be alleviated by introducing rate-compatible (or rateless) coding into the error correction scheme [181]. A key aspect of rateless coding is that the receiver is capable of decoding the original information, with high probability, even if only part of the original information is available. To this end, the transmitter creates a sequence of frames based on the original data, and the receiver combines the collected frames and decodes the original data once enough frames have been received.

(4) Adaptive optics: In adaptive optics (AO) a deformable mirror is used to pre-distort a transmitted laser beam so that on its passage through the turbulent atmosphere it arrives at the receiver as an undistorted wave. In most long range OW demonstrations that have been carried out to-date AO has been used to attempt to mitigate the effects of the turbulent channel. Unfortunately, as pointed out by Stotts et al. [164] "the ability of an adaptive optics (AO) system to compensate turbulence along a path is limited by the transmitter and receiver Rayleigh ranges, proportional to the diameter of the optics squared and inverse of the wavelength of light utilized." AO is ineffective when the total range is greater than the sum of the Rayleigh ranges. The Rayleigh range is $R_R = 0.7D^2/\lambda$, where D is the diameter of the transmitter or receiver aperture, and λ is the wavelength. For example with a 1.55-μm transmitter, and a 100-mm diameter receiver $R_R = 4.5$ km. An AO receiver can correct the wavefronts that actually arrive at a receiver aperture, for example to couple the light into single mode fiber. But, full reciprocal AO that couples transmitter and receiver is not possible unless the range is sufficiently short, or when the transmitter and

receiver are not moving fast laterally with respect to each other. Although AO has had considerable success in optical astronomy, its complexity, cost, and inability to correct for long range effects makes its widespread adoption for OW problematic. A full discussion of AO is beyond the scope of this article, but the study of Fried and co-workers [182–187] is particularly important in this context.

4.3. Satellite links and high-altitude platforms

Outdoor OW links are a high-bandwidth and low-power-consumption alternative in satellite communications [4] and OW links have been successfully demonstrated in satellite-to-satellite, earth-to-satellite and earth-to-moving-vehicle links [188–199]. Satellite systems are the most challenging category of outdoor OW systems, because of the very long distances that the optical beams propagate through vacuum or the atmosphere. Optical links that operate in vacuum (inter-satellite and space communications) are at least tens of thousands of kms long. Even if the vacuum is considered to be ideal (lossless), the associated path losses that increase with the square of the link distance are enormous, and vary from approximately 150 dB in satellite-to-satellite communications to approximately 230 dB in Mars-to-Earth communications. Earth-to-satellite optical links involves shorter distances, up to approximately 36,000 km for a geostationary orbit, but the total system losses are increased due to scattering and absorption in the atmosphere [150, 152]. Hence, a key consideration in satellite communications is the received optical power and several techniques have been proposed to mitigate the transmission associated losses. Optical amplification is an appealing candidate [200–203], especially in earth-to-satellite uplinks where low power consumption limitations do not apply. Another possible candidate in earth-to-satellite communications is the use of flying vehicles and high-altitude-platforms as beam relays [204–206]. When no amplification or intermediate relays are available, as in deep space communications [207], the losses may not be mitigated and more sophisticated receivers that operate at very low powers (photon-counting) are required [208].

A second important aspect of satellite systems is the selection of suitable modulation formats and coding schemes, depending on the application scope. Power efficiency is critical in satellite communications, since satellites recharge only by means of solar energy. Recently, a 5.6-Gb/s inter-satellite link demonstrated the suitability of binary phase shift keying (BPSK) for satellite OW links [192]. Deep space communications, on the other hand, are plagued by extreme path losses and favor the deployment of modulation formats with high peak-to-average power ratios, which ensure that energy peaks are detected. Multilevel PPM systems have been shown to be suitable for deployment in deep space communications [150, 209]. Efficient error-correcting coding is also required in deep space communications due to the extremely low powers received, and concatenated convolutional codes have being shown to perform close to the theoretical capacity limit [209]. In earth-to-satellite systems, coherent phase modulation schemes provide high sensitivity and the use of

BPSK has been previously reported on a similar 5.6 Gb/s 142 km OW link [210]. PPM has also been studied under the prism of providing sufficient peak power to penetrate through clouds [211, 212]. Error-correcting codes help to mitigate deep atmospheric fades, but there is a high possibility that the OW link will be lost due to adverse weather conditions altogether [213]. In such cases, temporal or spatial diversity is required to maintain a working earth-to-satellite optical link [214, 215].

A final challenge in satellite communications is beam acquisition, pointing, and tracking. Longer links typically require more accurate and sophisticated acquisition, pointing and tracking systems to maintain an operational OW link, since the slightest error is magnified by the link distance. From a communication perspective, pointing and tracking are not only important for maintaining the optical link, but also contribute to the link power penalty since misalignment caused by pointing and tracking errors result in power fluctuations at the receiver [216–218]. Pointing and tracking errors are caused mainly by (a) satellite vibrations [218], (b) errors in the calculation of the point-ahead angle [4], and (c) beam wander effects [219]. Pointing and tracking errors may be reduced by using advanced automated control methods in the pointing and tracking element [4], while their impact may be dampened by allowing adaptive variations in the transmitted power [218] or a broader optical beam [220]. In earth-to-satellite communications pointing and tracking also suffer from fades in the atmosphere, and it is possible that re-acquisition of the beam is required after a fade, leading to extended capacity loss. Reference [221] discusses how inertial sensors can be utilized to counter-effect beam loss due to atmospheric fading.

5. HYBRID OW/RF SYSTEMS

OW links can support high data rate communications in large unlicensed bandwidths, but their data rates and reliability are severely affected by atmospheric phenomena that include absorption, scattering and atmospheric turbulence [158]. While in clear conditions atmospheric attenuation for an OW link can be as low as a fraction of a few dB/km, heavy snow or fog can increase this attenuation to hundreds of dBs/km. Atmospheric turbulence, particularly in hot dry climates, causes signal fading that can degrade the link performance significantly. Besides, when the links involve mobility, maintaining accurate pointing and tracking over long distances becomes highly challenging, and link deterioration or complete link failure occurs. A key requirement is therefore to increase the reliability of OW links.

Improving the reliability of OW links, while still exploiting their high data carrying capabilities, is achieved by operating the OW link in conjunction with an RF link. Under this scheme of operation, the RF part may be used for link control, maintenance and backup functions that require relatively low line rates, while the broadband OW part transfers the bulk of data. This hybrid operation scheme improves reliability since RF links can be much less sensitive to atmospheric conditions and pointing errors, depending on their frequency of operation. In addition, when the OW links fail due to obscurations or fading, the RF links can act as a backup still providing a significant percentage of the OW system bandwidth. The RF data rate can be increased over a given bandwidth using higher order modulations, for instance quadrature amplitude modulation at the expense of transmitted power and system complexity. Even though RF frequencies have regulation restrictions and large bandwidths for high data rates at desired frequencies are difficult to obtain, unlicensed bands at 2.4, 5, and 60 GHz are available for the implementation of hybrid OW/RF systems.

The increased reliability of hybrid OW/RF over OW-only links enables them as a low-cost rapidly deployable last-mile communication technology. In last-mile applications, OW/RF can provide connectivity over longer distances within the eye safety limitations of laser radiation. In addition, OW/RF links can be used for both fixed as well as mobile multigigabit backbone networks [222–224]. Such networks may include ship-to-shore, ship-to-ship, ground-to-aircraft, and aircraft-to-aircraft links. The use of hybrid OW/RF systems for mobile robot team applications has been discussed in [225]. Taking advantage of the high bandwidths of OW links and reliability of RF links, such robot teams can transmit real-time video data in applications such as search and rescue operations. The use of hybrid OW/RF systems can also facilitate RF spectrum planning and utilization. If OW/RF hybrid systems are designed to meet a required quality of service then some of the existing RF-only links can be replaced with hybrid links. Hybrid links typically utilize only a fraction of the RF bandwidth in comparison with RF-only links [226]. The released RF spectrum can subsequently be made available for more critical applications.

5.1. Hybrid OW/RF channel model

In a hybrid OW/RF system, it is necessary to include both the OW and RF channel impairments in the overall system performance. In general, OW and RF links, depending on the frequency of operation, can be modeled independently to include the weather effects. As a result, separate channel models for the OW and RF links are mostly adequate to predict the hybrid system performance. Most practical OW systems employ intensity modulation with direct detection (IM/DD). The OW channel propagation results in a multiplicative effect on the transmitted optical intensity of the laser source with a channel gain factor. The channel gain includes atmospheric attenuation, pointing error effects and atmospheric turbulence. A detailed discussion of the channel impact on OW only systems has been presented in Section 4. Since the OW link operates with direct line-of-sight (LOS), the RF link too has LOS, and therefore the RF link can be modeled as a Ricean channel and often as an additive white Gaussian noise (AWGN) channel for good channel conditions. However, for longer and high altitude links severe fading in the RF link can occur. One example is [222], where the link operates over a distance of 150 km at a height of 10,000 ft. Both frequency-flat and frequency-selective fading were observed for a system operating in the X- and Ku-band at a data rate of 274 Mb/s.

When higher RF frequencies are considered (such as at millimeter wavelengths) the OW and RF channels may not be truly independent. Higher RF frequency bands support higher data rates, but they are also more affected by atmospheric effects and pointing errors. Consequently, higher RF frequencies are affected by scintillations, and moderate to heavy rain events were reported to affect both OW laser link at 850 nm and millimeter wave RF [227]. Pointing errors can also affect scintillation statistics for both channels [228]. In such a case, pointing error correlations between the two links cannot be ignored.

5.2. Hybrid OW/RF link architecture and networking

The RF link in hybrid OW/RF channels can be used for link acquisition, data transmission control and/or data transfer. Accordingly, hybrid systems are divided into three categories, depending on the usage of the RF link, as follows:

(1) Hybrid OW/RF using RF as the control link: The RF link is generally more reliable than the OW link and less sensitive to pointing. Thus, it can be used to assist in the pointing and acquisition of the OW link. It can also be used to provide packet retransmission requests. Successful RF retransmission requests can significantly improve system performance, as observed in [222] for a multi-Gb/s data rate system.

(2) Hybrid OW/RF with switching and data splitting: Since the OW link is the primary channel, the RF link can be used either as a backup or as a parallel link carrying a fraction of the transmitted data. An experimental 2.9 km hybrid OW/RF link is described in [229], where the RF link works in a standby mode. The laser link provides connection at 155 Mb/s and the received signal level is checked every 5 s. When the signal falls below a certain threshold, data transmission switches to the RF link, which operates at 2.4 GHz with a data rate of 11 Mb/s. Data transmission continues on the RF link until an improved signal level is detected and then the transmission switches back to the laser link. Although this hybrid OW/RF experiment is important, this approach under-utilizes resources, as only one link is active at any given time.

When using both OW and RF links concurrently, data is partitioned between the links. In [224], a dynamic load switching algorithm is employed and incremental load shifting from one link to the other is carried out, based on the status of the two links. For this purpose, BER is recorded on both links every minute and the averaging effect over a window is used for switching. A practical coded OW/RF switching scheme using an automatic load distribution between the links is described in [230]. The data packets are encoded using Raptor codes and sent over the two links at their respective rates of 1 Gb/s and 96 Mb/s. Both the links employ cyclic redundancy check (CRC) codes and discard corrupted packets. The receiver decodes the entire message when there are a sufficient number of correctly received packets. The transmitter continues sending encoded packets for a given message until it receives a 1-bit feedback message. Although for a given reliability level the data rate of this approach is less due to Raptor code overhead, the average data rate is improved by nearly six times in some cases. The use of Raptor codes for hybrid OW/RF systems is also considered in [231].

(3)Hybrid OW/RF with codeword partitioning: If the transmitted data are encoded for error correction, separate encoding and decoding over OW and RF links do not fully exploit the available diversity. For example, when the OW link deteriorates, a data packet encoded and decoded only over the OW link may still be received in error. If, however, some of the encoded bits are transmitted over a more reliable RF link, there is a higher probability of correctly receiving the packet. This idea has been recently explored by many researchers. In general, as more and more bits are transmitted over a more reliable RF channel, the hybrid link remains operational even with less available OW SNR. This gives rise to trade-offs between OW and RF link parameters.

A non-uniform rate-compatible low density parity check (LDPC) code is proposed in [232] for partitioning data into OW and RF links. The code rate is changed through puncturing of the coded bits. To get insights into the behavior of LDPC codes over hybrid channels, density evolution for hybrid channels is described in [226]. For a given code rate and OW/RF signal-to-noise ratios, the minimum RF data rate needed for the convergence of the decoder algorithm is obtained. This helps in the code requirements and design of hybrid systems. For example, the receiver can ask for an increase or decrease in RF rate in steps. Hybrid channel density evolution analysis for irregular LDPC codes shows that some further improvement in performance can be obtained. The performance improvement is higher for stronger turbulence channels. An experimental testbed verification demonstrating the joint decoding of LDPC codes over the two channels is carried out in [233]. Convolutional and turbo codes for hybrid channels are also investigated in the literature [234, 235]. Since these codes require simpler encoders than decoders, they can be useful at transmitters with limited resources. The convolutionally coded bits are partitioned into OW and RF channels using an optimized pattern and interleaved before modulation [234]. This bit-interleaved coded modulation is able to exploit the diversity of the two links. In [235], error analysis for turbo codes in hybrid channels showing trade-offs between RF and OW SNRs is described. Optimal turbo-coded bit allocations for puncturing and RF transmission are analyzed and related algorithms are also presented.

From a networking perspective, hybrid OW/RF networks are required to provide high-rate data connectivity with low end-to-end delay. A multihop network with short length links will provide high availability since the shorter OW links are less vulnerable to outages [224]. However, adding more hybrid nodes is costly and it also increases end-to-end delay. A hybrid wireless network of backbone nodes is considered in [223], where the nodes can be pointed to new nodes if necessary. Topology discovery, dissemination, link restoration, optimal OW/RF configurations and routing are very critical for network operation. These functions are accomplished through cooperation of the RF and OW subsystems. For link restoration, an increase in OW transmit power, bit rate reduction and hybrid link switching were explored. Alternatively, OW beam search along with RF connectivity can be used. If the OW link fails in a switched hybrid system then the connection switches to the RF link and optical beam probing can be continued until the OW link is restored [223]. A detailed OW/RF network analysis, with links using codeword partitioning, is an open research problem.

6. CONCLUSIONS AND FUTURE DIRECTIONS

This article has presented an overview of OW technologies, emphasizing their deployment in communication systems. Four distinct OW system categories (personal communication, indoor, outdoor, and hybrid OW/RF) have been discussed and key requirements and major applications areas have been identified for each category. Personal communication OW systems are required to provide high rate connectivity and very short connection establishment times. Both have been implemented in the IrDA-standardized physical layer and protocol stack. The current state of the art has enabled data transmission at Gb/s speeds and this technology is currently under standardization by the IrDA Giga-IR group. Future standardization efforts include a technology merge of infrared and USB technologies (IrUSB), as well as a comprehensive analysis and optimization of available protocols with the new Gb/s speeds in mind.

Indoor OW systems are required to allow for user mobility, which can be quite challenging due to the dispersive nature of the indoor channel. Dispersion mitigation is of critical importance in indoor OW and suitable mitigation techniques, including multi-spot diffusing links and coding, have been presented in this article. The deployment of OW has been limited in the past years, but recently VLCs have emerged as a future high-potential indoor OW technology. VLCs are being actively investigated by the VLCs Consortium and the IEEE 802.15.7 Task Group. The VLCs Consortium has already produced standards for VLCs and is currently cooperating with IrDA to promote OW systems for personal and indoor communications.

Outdoor OW systems are required to provide high data rates and remain functional at link distances that range from hundreds of meters to thousands of kms. Outdoor systems are seriously affected by atmospheric impairments, mainly obscuration and fading. These effects are increasingly deleterious as the link length increases and mitigation techniques like temporal/spatial diversity, coding/interleaving and aperture averaging are required to maintain link operation. At present, 10 Gb/s Ethernet compliant outdoor OW systems are commercially available for metro applications from several vendors. In the near future, link speeds are expected to rise, following the bandwidth increase in high-speed optical components that currently operate at 40-100 Gb/s.

Finally, hybrid OW/RF systems have been introduced to augment outdoor OW systems with increased reliability. The inclusion of the RF subsystem facilitates link establishment and operation, enhances error correction and data transfer control, and provides a backup link in the event of a failure. Hybrid systems are currently under research and researchers aim to identify which cooperation schemes between the OW and RF links provide the most reliable overall system operation.

COMPETING INTERESTS

The authors declare that they have no competing interests.

REFERENCES

1 Gfeller F, Bapst U: Wireless in-house data communication via diffuse infrared radiation. Proc IEEE 1979, 67(11):1474-1486.

2 Infrared Data Association (IrDA)[http://www.irda.org]

3 Visible Light Communications Consortium[http://www.vlcc.net]

4 Chan VWS: Optical satellite networks. IEEE/OSA J Lightw Technol 2003, 21(11):2811-2827. 10.1109/JLT.2003.819534

5 IEEE Standard for Wireless LAN Medium Access Control (MAC) and Physical Layer (PHY) Specifications. 1997.

6 IEEE 802.15 Working Group for WPAN[http://www.ieee802.org/15/]

7 IrDA: Advanced Infrared (AIr) MAC Draft Protocol Specification--Version 1.0, Infrared Data Association. 1999.

8 IrDA: Advanced Infrared Logical Link Control (AIrLC) Specification--Version 0.1, Infrared Data Association. 1999.

9 IrDA: Advanced Infrared Physical Layer Specification (AIr-PHY) --Version 1.0, Infrared Data Association. 1998.

10 IrDA: Point and Shoot Profile--Version 1.1, Infrared Data Association. 2000.

11 IrDA: Physical Layer Specification--Version 1.4, Infrared Data Association. 2001.

12 IrDA: Serial Infrared Physical Layer Specification--Giga-IR Addition Version 1.0, Infrared Data Association. 2009.

13 IEC 60825-1, Safety of Laser Products--Part 1: equipment classification, requirements, and user's guide, edition 1.2 International Electrotechnical Commission 2001.

14 IrDA: Serial Infrared Link Access Protocol (IrLAP) --Version 1.1, Infrared Data Association. 1996.

15 IrDA: Link Management Protocol (IrLMP)--Version 1.1, Infrared Data Association. 1996.

16 IrDA: Tiny TP: a flow-control mechanism for use with IrLMP--Version 1.1, Infrared Data Association. 1996.

17 IrDA: Object Exchange Protocol (OBEX™)--Version 1.5, Infrared Data Association. 2009.

18 IrDA: Link Access Protocol Specification for Giga-IR Addition--Version 1.0, Infrared Data Association. 2009.

19 IrDA: IrLAP Fast Connect (Application Note)--Version 1.0, Infrared Data Association. 2002.

20 IrDA: Minimal IrDA Protocol Implementation (IrDA Lite)--Version 1.0, Infrared Data Association. 1996.

21 IrDA: IrSimple (Infrared Simple) Profile--Version 1.01, Infrared Data Association. 2007.

22 IrDA: Serial Infrared Sequence Management Protocol for IrSimple--Version 1.0, Infrared Data Association. 2005.

23 IrDA: Infrared Financial Messaging (IrFM™) Point and Pay Profile--Version 1.0, Infrared Data Association. 2002.

24 IrDA: 'IrCOMM': Serial and Parallel Port Emulation over IR (Wire Replacement)--Version 1.0, Infrared Data Association. 1995.

25 IrDA: Infrared Universal Bus Specification--Version 0.11, Infrared Data Association. 2010.

26 Boucouvalas AC: Indoor ambient light noise and its effect on wireless optical links. IEE Proc Optoelectron 1996, 143(6):334-338. 10.1049/ip-opt:19960881

27 Kahn J, Barry J: Wireless infrared communications. Proc IEEE 1997, 85(2):265-298. 10.1109/5.554222

28 Kahn JM, Krause WJ, Carruthers JB: Experimental characterization of non-directed indoor infrared channels. IEEE Trans Commun 1995, 43(234):1613-1623.

29 Pérez-Jiménez R, Rabadán JA, López-Hernández FJ: Filtered modulation schemes for short distance infrared wireless communications. IEEE Trans Consumer Electron 2000, 46(2):275-282. 10.1109/30.846658

30 Valadas RT, Tavares AR, de Oliveira Duarte AM, Moreira AC, Lomba CT: The infrared physical layer of the IEEE 802.11 Standard for wireless local area networks. IEEE Commun Mag 1998, 36(12):107-112. 10.1109/35.735887

31 Shiu D, Kahn JM: Differential pulse position modulation for power-efficient optical communication. IEEE Trans Commun 1999, 47(8):1201-1210. 10.1109/26.780456

32 Ghassemlooy Z, Hayes AR: Digital pulse interval modulation for IR communication systems--a review. Int J Commun Sys 2000, 13(7-8):519-536. 10.1002/1099-1131(200011/12)13:7/8<519::AID-DAC454>3.0.CO;2-5MATH

33 Hirt W, Hassner M, Heise N: IrDA-VFIr (16 Mb/s): modulation code and system design. IEEE Personal Commun 2001, 8(1):58-71. 10.1109/98.904900

34 Widmer AX, Franaszek PA: A DC-balanced, partitioned-block, 8B/10B transmission code. IBM J Res Dev 1983, 27(5):440-451.

35 Street AM, Samaras K, O'Brien DC, Edwards DJ: Closed form expressions for baseline wander effects in wireless IR applications. IEE Electron Lett 1997, 33(12):1060-1062. 10.1049/el:19970739

36 Ghassemlooy Z: Investigation of the baseline wander effect on indoor optical wireless system employing digital pulse interval modulation. IET Commun 2008, 2(1):53-60. 10.1049/iet-com:20050552

37 Carruthers JB: JM Kahn, Multiple-subcarrier modulation for non-directed wireless infrared communication. IEEE J Sel Areas Commun 1996, 14(3):538-546. 10.1109/49.490239

38 You R, Kahn JM: Average power reduction techniques for multiple-subcarrier intensity-modulated optical signals. IEEE Trans Commun 2001, 49(12):2164-2171. 10.1109/26.974263MATH

39 Gonzalez O, Perez-Jimenez R, Rodriguez S, Rabadan J, Ayala A: OFDM over indoor wireless optical channel. IEE Proc Optoelectron 2005, 152(4):199-204. 10.1049/ip-opt:20045065

40 Mesleh R, Elgala H, Haas H: On the performance of different OFDM based optical wireless communication systems. OSA J Opt Commun Netw 2011, 3: 620-628.

41 Armstrong J: OFDM for optical communications. IEEE/OSA J Lightw Technol 2009, 27(3):189-2041.

42 Chatzimisios P, Boucouvalas AC: Packet delay analysis of the advanced infrared (AIr) CSMA/CA MAC protocol in optical wireless LANs. Int J Commun Syst 2005, 18(3):307-331. 10.1002/dac.705

43 Vitsas V, Boucouvalas AC: Performance analysis of the advanced infrared (AIr) CSMA/CA MAC protocol for wireless LANs. Wirel Netw 2003, 9(5):495-507. 10.1023/A:1024692217327

44 Barker P, Vitsas V, Boucouvalas AC: Simulation analysis of advanced infrared (Alr) MAC wireless communications protocol. IEE Proc Circ Dev Syst 2002, 149(3):193-197. 10.1049/ip-cds:20020444

45 Bianchi G: Performance analysis of the IEEE 802.11 distributed coordination function. IEEE J Sel Areas Commun 2000, 18(3):535-547. 10.1109/49.840210

46 Crow BP, Widjaja I, Kim LG, Sakai PT: IEEE 802.11 wireless local area networks. IEEE Commun Mag 1997, 35(9):116-126. 10.1109/35.620533

47 Huang P, Chatzimisios P, Boucouvalas A: Optimizing IrDA throughput by including processing time with physical layer consideration. OSA J Opt Netw 2005, 4: 323-331. 10.1364/JON.4.000323

48 Vitsas V, Boucouvalas AC: Optimization of IrDA IrLAP link access protocol. IEEE Trans Wirel Commun 2003, 2(5):926-938. 10.1109/TWC.2003.816776

49 Boucouvalas AC, Huang Pi: Modelling and optimising TinyTP over IrDA stacks. EURASIP J Wirel Commun Netw 2005, 2005(1):45-56.MATH

50 Boucouvalas AC, Huang Pi: OBEX over IrDA: performance analysis and optimization by considering multiple applications. IEEE/ACM Trans Netw 2006, 14(6):1292-1301.

51 Deccio CT, Ekstrom J, Partridge DR, Tew KB, Knutson CD: A study of the suitability of IrOBEX for high-speed exchange of large data objects. In Proc IEEE GLOBECOM. Volume 5. San Francisco; 2003:2664-2668.

52 Shah AM, Ara SS, Kitazumi G, Matsumoto M: IrSimple Modeling and Performance Evaluation for High-Speed Infrared Communications. In Proc IEEE GLOBECOM. San Francisco; 2006:1-6.

53 IrDA: Burst Data Protocol (IrBurst)--Version 1.0, Infrared Data Association. 2004.

54 Alam MS, Shawkat SA, Gontaro K, Mitsuji M: IrBurst modeling and performance evaluation for large data block exchange over high-speed IrDA links. IEICE Trans Commun 2008, 91(1):274-285.

55 Huang P, Boucouvalas A: IrBurst modelling and performance analysis in the presence of transmission errors. Wirel Personal Commun 2007, 41(1):111-125. 10.1007/s11277-006-9131-9

56 Huang P, Boucouvalas A: Future personal “e-payment”: IRFM. IEEE Wirel Commun 2006, 13(1):60-66. 10.1109/MWC.2006.1593526

57 Chu T-S, Gans M: High speed infrared local wireless communication. IEEE Commun Mag 1987, 25(8):4-10.

58 Heatley DJT, Wisely DR, Neild I, Cochrane P: Optical wireless: the story so far. IEEE Commun Mag 1998, 36(12):72-74. 79-82 10.1109/35.735881

59 Pang G, Kwan T, Liu H, Chan C-H: LED wireless. IEEE Ind Appl Mag 2002, 8(1):21-28. 10.1109/2943.974354

60 Hranilovic S: Wireless Optical Communication Systems. Springer, New York, NY; 2004.

61 Barry JR, Kahn JM: Link design for nondirected wireless infrared communications. OSA Appl Opt 1995, 34: 3764-3776.

62 Wang K, Nirmalathas A, Lim C, Skafidas E: 4 × 12.5 Gb/s WDM optical wireless communication system for indoor applications. IEEE/OSA J Lightw Technol 2011, 29(13):1988-1996.

63 Pakravan MR, Simova E, Kavehrad M: Holographic diffusers for indoor infrared communication systems. Int J Wirel Inf Netw 1997, 4(4):259-274. 10.1023/A:1018876326494

64 Elgala H, Mesleh R, Haas H: Indoor broadcasting via white LEDs and OFDM. IEEE Trans Consumer Electron 2009, 55(3):1127-1134.

65 Jungnickel V, Forck A, Haustein T, Kruger U, Pohl V, von Helmolt C: Electronic tracking for wireless infrared communications. IEEE Trans Wirel Commun 2003, 2(5):989-999. 10.1109/TWC.2003.817419

66 Plant DV, Kirk AG: Optical interconnects at the chip and board level: challenges and solutions. Proc IEEE 2000, 88(6):806-818. 10.1109/5.867694

67 Kim G, Han X, Chen R: An 8-Gb/s optical backplane bus based on microchannel interconnects: design, fabrication, and performance measurements. IEEE/OSA J Lightw Technol 2000, 18(11):1477-1486. 10.1109/50.896207

68 Bevan MG, Darrin MAG, Walts SC, Schneider W, Mills CS, Conde RF: Free-space optical data bus for spacecraft. In Proceedings of the 3rd annual Earth Science Technology Conference. Maryland, USA; 2003.

69 O'Brien DC, Faulkner GE, Zyambo EB, Jim K, Edwards DJ, Stavrinou P, Parry G, Bellon J, Sibley MJ, Lalithambika VA, Joyner VM, Samsudin RJ, Holburn DM, Mears RJ: Integrated transceivers for optical wireless communications. IEEE J Sel Topics Quantum Electron 2005, 11(1):173-183.

70 Miller D: Device requirements for optical interconnects to silicon chips. Proc IEEE 2009, 97(7):1166-1185.

71 Shelby RM, Hoffnagle JA, Burr GW, Jefferson CM, Bernal M-P, Coufal H, Grygier RK, Gunther H, Macfarlane RM, Sincerbox GT: Pixel-matched holographic data storage with megabit pages. OSA Opt Lett 1997, 22(19):1509-1511.

72 Burr GW, Ashley J, Coufal H, Grygier RK, Hoffnagle JA, Jefferson CM, Marcus B: Modulation coding for pixel-matched holographic data storage. OSA Opt Lett 1997, 22(9):639-641.

73 Bisaillon E, Brosseau DF, Yamamoto T, Mony M, Bernier E, Goodwill D, Plant DV, Kirk AG: Free-space optical link with spatial redundancy for misalignment tolerance. IEEE Photon Technol Lett 2002, 14(2):242-244.

74 Hranilovic S, Kschischang FR: Short-range wireless optical communication using pixelated transmitters and imaging receivers. In Proceedings of the IEEE International Conference on Communications. Volume 2. Paris, France; 2004:891-895.

75 Hranilovic S, Kschischang FR: A pixelated MIMO wireless optical communication system. IEEE J Sel Topics Quantum Electron 2006, 12(4):859-874.

76 Mesleh R, Elgala H, Haas H: Optical spatial modulation. OSA J Opt Commun Netw 2011, 3(3):234-244.

77 Takase D, Ohtsuki T: Spatial multiplexing in optical wireless MIMO communications over indoor environment. IEICE Trans Commun 2006, E89-B(4):1364-1370. 10.1093/ietcom/e89-b.4.1364

78 Villan R, Voloshynovskiy S, Koval O, Pun T: Multilevel 2-D bar codes: toward high-capacity storage modules for multimedia security and management. IEEE Trans Inf Foren Sec 2006, 1(4):405-420.

79 Mohamed MDA, Hranilovic S: Two-dimensional binary halftoned optical intensity channels. IET Commun 2008, 2(1):11-17. 10.1049/iet-com:20060385

80 Siegel PH: Information-theoretic limits of two-dimensional optical recording channels. In Proceedings of the IEEE Optical Data Storage Topical Meeting. Montreal, QC, Canada; 2006:165-167.

81 Huang L, Mathew G, Chong TC: Channel modeling and target design for two-dimensional optical storage systems. IEEE Trans Mag 2005, 41(8):2414-2424.

82 Joseph J, Waldman DA: Homogenized Fourier transform holographic data storage using phase spatial light modulators and methods for recovery of data from the phase image. Appl Opt 2006, 45: 6374-6380. 10.1364/AO.45.006374

83 Pu A, Psaltis D: High-density recording in photopolymer-based holographic three-dimensional disks. OSA Appl Opt 1996, 35(14):2389-2398.

84 Takase D, Ohtsuki T: Optical wireless MIMO communications (OMIMO). In IEEE Global Telecom Conf. Volume 2. Dallas; 2004:928-932.

85 Faucher J, Venditti MB, Plant DV: Application of parallel forward-error correction in two-dimensional optical-data links. IEEE/OSA J Lightw Technol 2003, 21(2):466-475. 10.1109/JLT.2003.808762

86 Ohbuchi E, Hanaizumi H, Hock LA: Barcode readers using the camera device in mobile phones. In Int Conference on Cyberworlds. Tokyo, Japan; 2004:260-265.

87 Kahn JM, Barry JR, Audeh MD, Carruthers JB, Krause WJ, Marsh GW: Non-directed infrared links for high-capacity wireless LANs. IEEE Personal Commun 1994, 1(2):12-25.

88 Kotzin M, van den Heuvel A: A duplex infra-red system for inbuilding communications. In Proceedings of the 36th IEEE Vehicular Technology Conference. Volume 36. Dallas, Texas; 1986:179-185.

89 Barry JR, Kahn JM, Krause WJ, Lee EA, Messerschmitt DG: Simulation of multipath impulse response for indoor wireless optical channels. IEEE J Sel Areas Commun 1993, 11(3):367-379. 10.1109/49.219552

90 Alqudah YA, Kavehrad M: MIMO characterization of indoor wireless optical link using a diffuse-transmission configuration. IEEE Trans Commun 2003, 51(9):1554-1560. 10.1109/TCOMM.2003.816945

91 Hashemi H, Yun G, Kavehrad M, Behbahani F, Galko PA: Indoor propagation measurements at infrared frequencies for wireless local area networks applications. IEEE Trans Veh Technol 1994, 43(3):562-576. 10.1109/25.312790

92 Pakravan MR, Kavehrad M: Indoor wireless infrared channel characterization by measurements. IEEE Trans Veh Technol 2001, 50(4):1053-1073. 10.1109/25.938580

93 Carruthers JB, Kahn JM: Modeling of nondirected wireless infrared channels. IEEE Trans Commun 1997, 45(10):1260-1268. 10.1109/26.634690

94 Jungnickel V, Pohl V, Nonnig S, von Helmolt C: A physical model of the wireless infrared communication channel. IEEE J Sel Areas Commun 2002, 20(3):631-640. 10.1109/49.995522

95 Lee H-S: A photon modeling method for the characterization of indoor optical wireless communication. Prog Electromagn Res PIER 2009, 92: 121-136.

96 Lopez-Hernandez FJ, Perez-Jimenez R, Santamarıa A: Ray-tracing algorithms for fast calculation of the channel impulse response on diffuse IR wireless indoor channels. Opt Eng 2000, 39(10):2775-2780. 10.1117/1.1287397

97 Gonzalez O, Rodriguez S, Perez-Jimenez R, Mendoza BR, Ayala A: Error analysis of the simulated impulse response on indoor wireless optical channels using a Monte Carlo-based ray-tracing algorithm. IEEE Trans Commun 2005, 53(1):124-130. 10.1109/TCOMM.2004.840625

98 González O, Rodríguez S, Pérez-Jiménez R, Mendoza BR, Ayala A: Comparison of Monte Carlo ray-tracing and photon-tracing methods for calculation of the impulse response on indoor wireless optical channels. OSA Opt Express 2011, 19: 1997-2005.

99 Cocheril Y, Vauzelle R: A new ray-tracing based wave propagation model including rough surfaces scattering. Prog Electromagn Res PIER 2007, 75: 357-381.

100 Audeh MD, Kahn JM, Barry JR: Performance of pulse position modulation on measured non-directed indoor infrared channels. IEEE Trans Commun 1996, 44(6):654-659. 10.1109/26.506380

101 Audeh MD, Kahn JM, Barry JR: Decision-feedback equalization of pulse-position modulation on measured nondirected indoor infrared channels. IEEE Trans Commun 1999, 47(4):500-503. 10.1109/26.764921

102 Lee DC, Kahn JM, Audeh MD: Trellis-coded pulse position modulation for indoor wireless infrared communications. IEEE Trans Commun 1997, 45(9):1080-1087. 10.1109/26.623072

103 Lee DC, Kahn JM: Coding and equalization for PPM on wireless infrared channels. IEEE Trans Commun 1999, 47(2):255-260. 10.1109/26.752131

104 Gfeller F, Hirt W: Advanced infrared (AIr): physical layer for reliable transmission and medium access. In Proceedings of the International Zurich Seminar on Broadband Communications. Zurich, Switzerland; 2000:77-84.

105 Lee D, Kahn J: Experimental 25-Mb/s wireless infrared link using 4-PPM with scalar decision-feedback equalization. In Proceedings of the IEEE International Conference on Communications. Volume 1. Atlanta, GA, USA; 1998:26-30.

106 Marsh GW, Kahn JM: 50-Mb/s diffuse infrared free-space link using on-off keying with decision-feedback equalization. IEEE Photon Technol Lett 1994, 6(10):1268-1270.

107 Marsh GW, Kahn JM: Performance evaluation of experimental 50-Mb/s diffuse infrared wireless link using on-off keying with decision-feedback equalization. IEEE Trans Commun 1996, 44(11):1496-1504. 10.1109/26.544466

108 Yun G, Kavehrad M: Indoor infrared wireless communications using spot diffusing and fly-eye receivers. Can J Electron Comput E 1993, 18(4):151-157.

109 Alsaadi FE, Elmirghani JMH: Mobile multigigabit indoor optical wireless systems employing multibeam power adaptation and imaging diversity receivers. OSA J Opt Commun Netw 2011, 3: 27-39.

110 Alsaadi FE, Elmirghani JMH: High-speed spot diffusing mobile optical wireless system employing beam angle and power adaptation and imaging receivers. IEEE/OSA J Lightw Technol 2010, 28(16):2191-2206.

111 Jivkova ST, Hristov BA, Kavehrad M: Power-efficient multispot-diffuse multiple-input-multiple-output approach to broad-band optical wireless communications. IEEE Trans Veh Technol 2004, 53(3):882-889. 10.1109/TVT.2004.825779

112 Jivkova ST, Kavehrad M: Multispot diffusing configuration for wireless infrared access. IEEE Trans Commun 2000, 48(6):970-978. 10.1109/26.848558

113 Jivkova ST, Kavehrad M: Holographic optical receiver front end for wireless infrared indoor communications. OSA Appl Opt 2001, 40: 2828-2835.

114 Kahn JM, You R, Djahani P, Weisbin AG, Teik BK, Tang A: Imaging diversity receivers for high-speed infrared wireless communication. IEEE Commun Mag Vol 1998, 36: 88-94.

115 Carruthers J, Kahn JM: Angle diversity for nondirected wireless infrared communication. IEEE Trans Commun 2000, 48(6):960-969. 10.1109/26.848557

116 Jivkova S, Kavehrad M: Receiver designs and channel characterization for multi-spot high-bit-rate wireless infrared communications. IEEE Trans Commun 2001, 49(12):2145-2153. 10.1109/26.974261

117 Kavehrad M, Jivkova S: Indoor broadband optical wireless communications: optical subsystems design and their impact on channel characteristics. IEEE Wirel Commun Mag Vol 2003, 10: 30-35.

118 Al-Ghamdi AG, Elmirghani JMH: Line strip spot-diffusing transmitter configuration for optical wireless systems influenced by background noise and multipath dispersion. IEEE Trans Commun 2004, 52(1):37-45. 10.1109/TCOMM.2003.822160

119 Al-Ghamdi AG, Elmirghani JMH: Multiple spot diffusing geometries for indoor optical wireless communication systems. Int J Commun Syst 2003, 16: 909-922. 10.1002/dac.627

120 Jungnickel V, von Helmolt C, Kruger U: Broadband wireless infrared LAN architecture compatible with ethernet protocol. Electron Lett 1998, 34(25):2371-2372. 10.1049/el:19981676

121 Narasimhan R, Audeh MD, Kahn JM: Effect of electronic-ballast fluorescent lighting on wireless infrared links. IEE Proc Optoelectron 1996, 143(6):347-354. 10.1049/ip-opt:19960877

122 Moreira AJC, Valadas RT, de Olveira Duarte AM: Optical interference produced by artificial light. Wirel Netw 1997, 3(2):131-140. 10.1023/A:1019140814049

123 Fernando XN, Krishnan S, Sun H, Kamyar KM: Adaptive denoising at infrared wireless receivers. Proc SPIE 2003, 5074: 199-207.

124 Rajbhandari S, Ghassemlooy Z, Angelova M: Effective denoising and adaptive equalization of indoor optical wireless channel with artificial light using the discrete wavelet transform and artificial neural network. IEEE/OSA J Lightw Technol 2009, 27(20):4493-4500.

125 Rajbhandari S, Ghassemlooy Z, Angelova M: Wavelet--artificial neural network receiver for indoor optical wireless communications. IEEE/OSA J Lightw Technol 2011, 29(17):2651-2659.

126 Wong KK, O'Farrell T: Spread spectrum techniques for indoor wireless IR communications. IEEE Wirel Commun 2003, 10(2):54-63. 10.1109/MWC.2003.1196403

127 Komine T, Nakagawa M: Fundamental analysis for visible-light communication system using LED lights. IEEE Trans Consumer Electron 2004, 50(1):100-107. 10.1109/TCE.2004.1277847

128 O'Brien D, Katz M: Optical wireless communications within fourth-generation wireless systems [invited]. OSA J Opt Netw 2005, 4(6):312-322. 10.1364/JON.4.000312

129 Japan Electronics and Information Technology Industries Association[http://www.jeita.or.jp]

130 OMEGA, the Home Gigabit Access Project[http://www.ict-omega.eu/]

131 The OMEGA Project Flyer [Online]. Available: http://www.ict-omega.eu/fileadmin/documents/flyers/OMEGA_Project_Flyer.pdf

132 Future networks, Projects Portfolio, Seventh EU Framework Programme for Research and Technological Development, Information Society and Media, European Commission 2008.

133 IEEE 802.15 WPAN-Visible Light Communication Interest Group (IGvlc)[http://ieee802.org/15/pub/IGvlc.html]

134 Sugiyama H, Haruyama S, Nakagawa M: Experimental investigation of modulation method for visible-light communications. IEICE Trans Commun 2006, E89-B(12):3393-3400. 10.1093/ietcom/e89-b.12.3393

135 Elgala H, Mesleh R, Haas H, Pricope B: OFDM visible light wireless communication based on white LEDs. In Proc IEEE Vehicular Technol Conf. Volume 4212879. Dublin, Ireland; 2007:2185-2189.

136 Grubor J, Randel S, Langer K-D, Walewski JW: Broadband information broadcasting using LED-based interior lighting. IEEE/OSA J Lightw Technol 2008, 26(24):3883-3892.

137 Zeng L, O'Brien DC, Minh HL, Faulkner GE, Lee K, Jung D, Oh Y, Won T: High data rate multiple input multiple output (MIMO) optical wireless communications using white led lighting. IEEE J Sel Areas Commun 2009, 27(9):1654-1662.

138 Kwon JK: Inverse source coding for dimming in visible light communications using NRZ-OOK on reliable links. IEEE Photon Technol Lett 2010, 22(19):1455-1457.

139 Sunghwan K, Sung-Yoon J: Novel FEC coding scheme for dimmable visible light communication based on the modified Reed-Muller codes. IEEE Photon Technol Lett 2011, 23(20):1514-1516.

140 Komine T, Lee JH, Haruyama S, Nakagawa M: Adaptive equalization system for visible light wireless communication utilizing multiple white led lighting equipment. IEEE Trans Wirel Commun 2009, 8(6):2892-2900.

141 Minh HL, O'Brien D, Faulkner G, Zeng L, Lee K, Jung D, Oh Y: 80 Mbit/s visible light communications using pre-equalized white LED. In Proc Eur Conf Opt Commun no. P.6.06. Brussels, Belgium; 2008:1-2.

142 Minh HL, O'Brien D, Faulkner G, Zeng L, Lee K, Jung D, Oh Y, Won ET: 100-Mb/s NRZ visible light communications using a postequalized white LED. IEEE Photon Technol Lett 2009, 21(15):1063-1065.

143 Grubor J, Lee SCJ, Langer K-D, Koonen T, Walewski JW: Wireless high-speed data transmission with phosphorescent white-light LEDs. In Proc Eur Conf Opt Commun, PD3.6. Berlin, Germany; 2007:1-2.

144 Vučić J, Kottke C, Nerreter S, Langer K-D, Walewski JW: 513 Mbit/s visible light communications link based on DMT-modulation of a white LED. IEEE/OSA J Lightw Technol 2010, 28(24):3512-3518.

145 Vučić J, Kottke C, Nerreter S, Büttner A, Langer K-D, Walewski JW: White light wireless transmission at 200+ Mb/s net data rate by use of

discrete-multitone modulation. IEEE Photon Technol Lett 2009, 21(20):1511-1513.

146 Lee K, Park H: Modulations for visible light communications with dimming control. IEEE Photon Technol Lett 2011, 23(16):1136-1138.

147 IEEE 802.15.7 VLC PHY/MAC Proposal[https://mentor.ieee.org/802.15/dcn/09/15-09-0733-00-0007-vlc-phy-mac-proposal-samsung-etri.pdf]

148 Ntogari G, Kamalakis T, Walewski JW, Sphicopoulos T: Combining illumination dimming based on pulse-width modulation with visible-light communications based on discrete multitone. OSA J Opt Commun Netw 2011, 3: 56-65.

149 Liu Q, Qiao C, Mitchell G, Stanton S: Optical wireless communication networks for first- and last-mile broadband access [Invited]. OSA J Opt Netw 2005, 4: 807-828. 10.1364/JON.4.000807

150 Chan VWS: Free-space optical communications. IEEE/OSA J Lightw Technol 2006, 24(12):4750-4762.

151 Zilberman A, Kopeika NS: Laser beam wander in the atmosphere: implications for optical turbulence vertical profile sensing with imaging LIDAR. J Appl Remote Sens 2008., 2(023540):

152 Andrews LC, Phillips RL: Laser Beam Propagation Through Random Media. 2nd edition. SPIE Press, Bellingham, WA; 2005.

153 Tatarski VI: Wave Propagation in a Turbulent Medium. McGraw-Hill, New York; 1961. (Reprinted by Dover, New York, 1967)MATH

154 Chernov LA: Wave Propagation in a Random Medium. McGraw-Hill, New York; 1960. (Reprinted by Dover, New York, 1967)

155 Ishimaru AK: Wave Propagation and Scattering in Random Media. Volume I and II. Academic Press, New York; 1978. Vol. II Chapters 16-20 are especially relevant

156 Strohbehn JW: Laser Beam Propagation Through the Atmosphere. Springer-Verlag, New York; 1978.

157 Zuev VE: Laser Beams in the Atmosphere. Consultants Bureau, New York; 1982.

158 Andrews LC, Phillips RL, Hopen CY: Laser Beam Scintillation with Applications. SPIE Press, Bellingham, WA; 2001.

159 Lawrence RS, Strohbehn JW: A survey of clear-air propagation effects relevant to optical communications. Proc IEEE 1970, 58: 1523-1545.

160 Fante RW: Electromagnetic beam propagation in turbulent media. Proc IEEE 1975, 63: 1669-1692.

161 Yura HT, McKinley WG: Optical scintillation statistics for IR ground-to-space laser communication systems. OSA Appl Opt 1983, 22: 3353-3358.

162 Al-Habash MA, Andrews LC, Phillips RL: Mathematical model for the irradiance probability density function of a laser beam propagating through turbulent media. Opt Eng 2001, 40(8):1554-1562. 10.1117/1.1386641

163 Anguita JA, Djordjevic IB, Neifeld MA, Vasic BA: Shannon capacities and error-correction codes for optical atmospheric turbulence channels. OSA J Opt Netw 2005, 4(9):586-601. 10.1364/JON.4.000586

164 Stotts LB, Kolodzy P, Pike A, Graves B, Daugherty D, Douglas J: Free space optical communication link budget estimation. OSA Appl Opt 2010, 49(28):5333-5343.

165 Arnon S: Effects of atmospheric turbulence and building sway on optical wireless-communication systems. OSA Opt Lett 2003, 28: 129-131.

166 Kedar D, Arnon S: Optical wireless communication through fog in the presence of pointing errors. OSA Appl Opt 2003, 42: 4946-4954.

167 Yuksel H, Milner S, Davis CC: Aperture averaging for optimizing receiver design and system performance on free-space optical communication links. OSA J Opt Netw 2005, 4(8):462-475. 10.1364/JON.4.000462

168 Yuksel H, Davis CC: Aperture averaging analysis and aperture shape invariance of received scintillation in free space optical communication links. In Proc SPIE Free-Space Laser Commun VI. Volume 6304. San Diego; 2006.

169 Wilson SG, Brandt-Pearce M, Cao Q, Baedke M: Optical repetition MIMO transmission with multi-pulse PPM. IEEE J Sel Areas Commun 2005, 23(9):1901-1910.

170 Navidpour SM, Uysal M, Kavehrad M: BER performance of free-space optical transmission with spatial diversity. IEEE Trans Wirel Commun 2007, 6(8):2813-2819.

171 Goldsmith A, Jafar SA, Jindal N, Vishwanath S: Capacity limits of MIMO channels. IEEE J Sel Areas Commun 2003, 21(5):684-702. 10.1109/JSAC.2003.810294

172 Khalighi M-A, Schwartz N, Aitamer N, Bourennane S: Fading reduction by aperture averaging and spatial diversity in optical wireless systems. OSA J Opt Commun Netw 2009, 1: 580-593.

173 García-Zambrana A, Castillo-Vázquez C, Castillo-Vázquez B: Outage performance of MIMO FSO links over strong turbulence and misalignment fading channels. OSA Opt Express 2011, 19: 13480-13496.

174 Davis CC, Smolyaninov II: The effect of atmospheric turbulence on bit-error-rate in an on-off-keyed optical wireless system. Proc SPIE Free-Space Laser Commun Laser Imag 2002, 4489: 126-137.

175 Trisno S, Smolyaninov II, Milner SD, Davis CC: Delayed diversity for fade resistance in optical wireless communication system through simulated turbulence. Proc SPIE Opt Transmission Syst Equip WDM Netw III 2005, 5596: 385-393. Philadelphia, Pennsylvania

176 Henniger H, Gonzalez A: Transmission scheme and error protection for simplex long-distance atmospheric FSO systems. Special Issue Mediterranean J Electron Commun Hybrid RF Opt Wirel Commun 2006, 2(3):118-126.

177 Zhu X, Kahn JM: Performance bounds for coded free-space optical communications through atmospheric turbulence channels. IEEE Trans Commun 2003, 51: 1233-1239. 10.1109/TCOMM.2003.815052

178 Djordjevic IB: LDPC-coded MIMO optical communication over the atmospheric turbulence channel using Q-ary pulse-position modulation. OSA Opt Express 2007, 15: 10026-10032.

179 Greco JA: Design of the high-speed framing, FEC, and interleaving hardware used in a 5.4 km free-space optical communication experiment. SPIE Proc Free-space Laser Communications IX 2009., 7464:

180 Henniger H, David F, Giggenbach D, Rapp C: Evaluation of FEC for the atmospheric optical IM/DD channel. SPIE Proc Free-Space Laser Communication Technologies XV 2003., 4975:

181 Anguita JA, Neifeld MA, Hildner B, Vasic B: Rateless coding on experimental temporally correlated FSO channels. IEEE/OSA J Lightw Technol 2010, 28(7):990-1002.

182 Fried DL: Limiting resolution looking down through atmosphere. JOSA 1966, 56: 1380-1384. 10.1364/JOSA.56.001380

183 Fried DL: Aperture averaging of scintillation. JOSA 1967, 57: 169-172. 10.1364/JOSA.57.000169

184 Fried DL, Seidman JB: Laser-beam scintillation in the atmosphere. JOSA 1967, 57: 181-185. 10.1364/JOSA.57.000181

185 Greenwood DP, Fried DL: Power spectra requirements for wave-front-compensative systems. JOSA 1976, 66: 193-206. 10.1364/JOSA.66.000193

186 Fried DL: Anisoplanatism in adaptive optics. JOSA 1982, 72: 52-61. 10.1364/JOSA.72.000052

187 Barchers JD, Fried DL: Optimal control of laser beams for propagation through a turbulent medium. JOSA 2002, A: 1779-1793.

188 Sodnik Z, Furch B, Lutz H: Optical intersatellite communication. IEEE J Sel Topics Quantum Electron 2010, 16(5):1051-1057.

189 Tolker-Nielsen T, Oppenhaeuser GT: In orbit test result of an operational optical intersatellite link between ARTEMIS and SPOT4, SILEX. Proc SPIE 2002., 6635(59430):

190 Jono T, Takayama Y, Kura N, Ohinata K, Koyama Y, Shiratama K, Sodnik Z, Demelenne B, Bird A, Arai K: OICETS on-orbit laser communication experiments. Proc SPIE 2006, 6105(610503):1-11.

191 Fujiwara Y, Jono T, Yamawaki T, Arai K, Toyoshima M, Kunimori H, Sodnik Z, Bird A, Demelenne B: Optical inter-orbit communications engineering test satellite (OICETS). Acta Astronautica 2007, 61(1-6):163-175. 10.1016/j.actaastro.2007.01.021

192 Smutny B, Kampfner H, Muhlnikel G, Sterr U, Wandernoth B, Heine F, Hildebrand U, Dallmann D, Reinhardt M, Freier A, Lange R, Bohmer K, Feldhaus T, Muller J, Weichert A, Seel S, Meyer R, Czichy R: 5.6 Gb/s optical intersatellite communication link. Proc SPIE 2009., 7199(719906):

193 Wilson KE, Lesh JR, Yan TY: GOPEX: a laser uplink to the Galileo spacecraft on its way to Jupiter. Proc SPIE 1993, 1866: 138-146.

194 Wilson KE, Lesh JR, Araki K, Arimoto Y: Overview of the ground-to-orbit lasercom demonstration (GOLD). Proc SPIE 1997, 2990: 23.

195 Toyoshima M, Yamakawa S, Yamawaki T, Arai K, Reyes M, Alonso A, Sodnik Z, Demelenne B: Ground-to-satellite optical link tests between Japanese laser communications terminal and European geostationary satellite ARTEMIS. Proc SPIE 2004, 5338: 1-15.

196 Horwath J, Perlot N, Knapek M, Moll F: Experimental verification of optical backhaul links for high-altitude platform networks: atmospheric turbulence and downlink availability. Int J Satell Commun Netw 2007, 25(5):501-528. 10.1002/sat.888

197 Perlot N, Knapek M, Giggenbach D, Horwath J, Brechtelsbauer M, Takayama Y, Jono T: Results of the optical downlink experiment KIODO from OICETS satellite to optical ground station oberpfaffenhofen (OGS-OP). Proc SPIE 2007., 6457(645704):

198 Horwath J, Fuchs C: Aircraft to ground unidirectional laser-comm terminal for high resolution sensors. Proc SPIE 2009., 7199(719909):

199 Baister G, Kudielka K, Dreischer T, Tüchler M: Results from the DOLCE (deep space optical link communications experiment) Project. Proc SPIE 2009., 7199(71990B):

200 Pe'er IL, Naftali N, Yogev A: High power, solar pumped, Nd:YAG, laser amplifier for free space laser communication. Proc SPIE 1997, 3139: 194-204.

201 Rochat E, Dändliker R, Haroud K, Czichy RH, Roth U, Costantini D, Holzner R: Fiber amplifiers for coherent space communication. IEEE J Sel Topics Quantum Electron 2001, 7(1):64-81. 10.1109/2944.924012

202 Polishuk A, Arnon S: Optimization of a laser satellite communication system with an optical preamplifier. J Opt Soc Am A 2004, 21(7):1307-1315. 10.1364/JOSAA.21.001307

203 Dawson JW, Messerly MJ, Beach RJ, MYSEA Stappaerts, Sridharan AK, Pax PH, Heebner JE, Siders CW, Barty CPJ: Analysis of the scalability of diffraction-limited fiber lasers and amplifiers to high average power. OSA Opt Express 2008, 16(17):13240-13266.

204 Djuknic GM, Freidenfelds J, Okunev Y: Establishing wireless communications services via high-altitude aeronautical platforms: a concept whose time has come? IEEE Commun. Mag 1997, 35(9):128-135.

205 Knapek M, Horwath J, Moll F, Epple B, Courville N, Bischl H, Giggenbach D: Optical high-capacity satellite downlinks via high-altitude platform relays. Proc SPIE 2007., 6709(67090E):

206 Fidler F, Knapek M, Horwath J, Leeb WR: Optical communications for high-altitude platforms. IEEE J Sel Topics Quantum Electron 2010, 16(5):1058-1070.

207 Boroson DM, Biswas A, Edwards BL: MLCD: Overview of NASA's Mars laser communications demonstration system. Proc SPIE 2004, 5338: 16-28.

208 Mendenhall JA, Candell LM, Hopman PI, Zogbi G, Boroson DM, Caplan DO, Digenis CJ, Hearn DR, Shoup RC: Design of an optical photon counting array receiver system for deep-space communications. Proc IEEE 2007, 95(10):2059-2069.

209 Moision B, Hamkins J: Deep-space optical communications downlink budget: modulation and coding. IPN Prog Rep 2003, 42-154: 1-28.

210 Lange R, Smutny B, Wandernoth B, Czichy R, Giggenbach D: 142 km, 5.625 Gb/s free-space optical link based on homodyne BPSK modulation. Proc SPIE 2006., 6105(61050A):

211 Muhammad SS, Javornik T, Jelovcan I, Ghassemlooy Z, Leitgeb E: Comparison of hard-decision and soft-decision channel coded M-ary PPM

performance over free space optical links. Eur Trans Telecommun 2008, 20(8):746-757.

212 Djordjevic IB, Vasic B, Neifeld MA: Multilevel coding in free-space optical MIMO transmission with Q-ary PPM over the atmospheric turbulence channel. IEEE Photon Technol Lett 2006, 18(14):1491-1493.

213 Toyoshima M, Takayama Y, Takahashi T, Suzuki K, Kimura S, Takizawa K, Kuri T, Klaus W, Toyoda M, Kunimori H, Jono T, Arai K: Ground-to-satellite laser communication experiments. IEEE Aerospace Electron Syst Mag 2008, 23(8):10-18.

214 Zhu X, Kahn JM: Free-space optical communication through atmospheric turbulence channels. IEEE Trans Commun 2002, 50(8):1293-1300. 10.1109/TCOMM.2002.800829

215 Lee J, Chan VWS: Diversity coherent and incoherent receivers for free-space optical communication in the presence and absence of interference. OSA J Opt Commun Netw 2009, 1(5):463-483.

216 Chen CC, Gardner CS: Impact of random pointing and tracking errors on the design of coherent and incoherent optical intersatellite communication links. IEEE Trans Commun 1989, 37(3):252-260. 10.1109/26.20099

217 Liu X: Optimal transmitter power of an intersatellite optical communication system with reciprocal Pareto fading. OSA Appl Opt 2010, 49: 915-919.

218 Arnon S, Kopeika NS, Kedar D, Zilberman A, Arbel D, Livne A, Guelman M, Orenstain M, Michalik H, Ginati A: Performance limitation of laser satellite communication due to vibrations and atmospheric turbulence: down-link scenario. Int J Satell Commun Netw 2003, 21(6):561-573. 10.1002/sat.769

219 Perlot N: Atmospheric occultation of optical intersatellite links: coherence loss and related parameters. OSA Appl Opt 2009, 48(12):2290-2302.

220 Arnon S, Rotman S, Kopeika NS: Beam width and transmitter power adaptive to tracking system performance for free-space optical communication. OSA Appl Opt 1997, 36(24):6095-6101.

221 Ortiz GG, Lee S, Monacos S, Wright M, Biswas A: Design and development of a robust ATP subsystem for the altair UAV-to-ground lasercomm 2.5 Gb/s demonstration. Proc SPIE 2003, 4975: 103-114.

222 Stotts LB, Andrews LC, Cherry PC, Foshee JJ, Kolodzy PJ, Mcintire WK, Northcott M, Phillips RL, Pike HA, Stadler B, Young DW: Hybrid optical RF airborne communications. Proc IEEE 2009, 97(6):1109-1127.

223 Milner SD, Davis CC: Hybrid free space optical/RF networks for tactical operations. In Proc IEEE Milcom 1. Monterey; 2004:409-415.

224 Izadpanah H, Elbatt T, Kukshya V, Dolezal F, Ryu BK: High-availability free space optical and RF hybrid wireless networks. IEEE Wirel Commun 2003, 10(2):45-53. 10.1109/MWC.2003.1196402

225 Derenick J, Thorne C, Spletzer S: On the deployment of a hybrid FSO/RF mobile ad-hoc network. In Proc IEEE/RSJ Int Conf Intelligent Robots and Syst. Edmonton; 2005:3990-3996.

226 Tapse H, Borah DK: Hybrid Optical/RF channels: characterization and performance study using low density parity check codes. IEEE Trans Commun 2009, 57: 3288-3297.

227 Kukshya V, Rappaport TS, Izadpanah H, Tangonan G, Guerrero RA, Mendoza JK, Lee B: Free-space optics and high-speed RF for next generation networks--propagation measurements. In Proc IEEE VTC-Fall 1. Vancouver; 2002:616-620.

228 Borah DK, Voelz DG: Pointing error effects on free space optical communication links in the presence of atmospheric turbulence. IEEE/OSA J Lightwave Technol 2009, 27: 3965-3973.

229 Akbulut A, Ilk HG, Ari F: Design, availability and reliability analysis on an experimental outdoor FSO/RF communication system. Proc Transparent Opt Netw 2005, 1: 403-406.

230 Zhang W, Hranilovic S, Shi C: Soft-switching hybrid FSO/RF links using short-length raptor codes: design and implementation. IEEE J Sel Areas Commun 2009, 27: 1698-1708.

231 Abdul Hussein A, Oka A, Nguyen T, Lampe L: Rateless coding for hybrid free-space optical and radio-frequency communication. IEEE Trans Wirel Commun 2010, 9: 907-913.

232 Vangala S, Pishro-Nik H: A highly reliable FSO/RF communication system using efficient codes. In Proc IEEE Globecom. Washington, D.C.; 2007:2232-2236.

233 Luna R, Borah DK, Jonnalagadda R, Voelz D: Experimental demonstration of a hybrid link for mitigating atmospheric turbulence effects in free space optical communication. IEEE Photon Technol Lett 2009, 21: 1196-1198.

234 He B, Schober R: Bit-interleaved coded modulation for hybrid RF/FSO systems. IEEE Trans Commun 2009, 57: 3753-3763.

235 Tapse H, Borah DK: J Perez-Ramirez, Hybrid optical/RF channel performance analysis for turbo codes. IEEE Trans Commun 2011, 59: 1389-1399.

CHAPTER 2

Evolutionary Algorithms for Wireless Communications — A Review of the State-of-the art

Sotirios K. Goudos[1]

[1] Radiocommunications Laboratory, Section of Applied and Environmental Physics, Department of Physics, Aristotle University of Thessaloniki, Thessaloniki, Greece

1. INTRODUCTION

Several evolutionary algorithms (EAs) have emerged in the past decade that mimic biological entities behavior and evolution. Darwin's theory of evolution is the major inspiration source for EAs. The foundation of Darwin's theory of evolution is natural selection. The study of evolutionary algorithms began in the 1960s. Several researchers independently developed three mainstream evolutionary algorithms, namely, genetic algorithms [1, 2], evolutionary programming [3], and evolution strategies [4]. EAs are widely used for the solution of single and multi-objective optimization problems. Swarm Intelligence (SI) algorithms are also a special type of EAs. SI can be defined as the collective behavior of decentralized and self-organized swarms. SI algorithms among others include Particle Swarm Optimization (PSO) [5], Ant Colony Optimization [6], and Artificial Bee Colony (ABC) [7].

PSO is an evolutionary algorithm that mimics the swarm behavior of bird flocking and fish schooling [5]. The most common PSO algorithms include the classical Inertia Weight PSO (IWPSO) and Constriction Factor PSO (CFPSO) [8]. PSO is an easy to implement algorithm with computational efficiency. The

PSO algorithm is inherently used only for real-valued problems. An option to expand PSO for discrete valued problems also exists. Among others PSO algorithms include, the barebones (BB) and the exploiting barebones (BBExp). BBPSO has been successfully applied to the cell to switch assignment problem [9].

Artificial Bee Colony (ABC) [7] is a recently proposed SI algorithm, which has been applied to several real world engineering problems. The ABC algorithm models and simulates the honey bee behavior in food foraging. In the ABC algorithm, a potential solution to the optimization problem is represented by the position of a food source while the food source corresponds to the quality (objective function fitness) of the associated solution. The ABC algorithm has been successfully applied to several problems in wireless communications [10]. ABC variants that improve the original algorithm have also been proposed [11].

Ant Colony Optimization (ACO) is a population-based metaheuristic introduced by Marco Dorigo [12]. This algorithm was inspired by the behaviour of real ants. The algorithm is based on the fact that ant colonies can find the shortest path between their nest and a food source just by depositing and reacting to pheromones while they are exploring their environment. ACO is suitable for solving combinatorial optimization problems, which are common in wireless communications.

Differential evolution (DE) [13, 14] is a population-based stochastic global optimization algorithm, which has been used in several real world engineering problems. Several DE variants or strategies exist. One of the DE advantages is that very few control parameters have to be adjusted in each algorithm run. However, the control parameters involved in DE are highly dependent on the optimization problem. Moreover, the selection of the appropriate strategy for trial vector generation requires additional computational time using a trial-and-error search procedure. Therefore, it is not always an easy task to fine-tune the control parameters and strategy. Since finding the suitable control parameter values and strategy in such a way is often very time-consuming, there has been an increasing interest among researchers in designing new adaptive and self-adaptive DE variants. Self adaptive DE (SaDE), is a DE algorithm that self-adapts both control parameters and strategy based on learning experiences from previous generations is presented in [15-17]. SaDE has been applied to microwave filter design, [18], and to antenna arrays synthesis [19].

The purpose of this chapter is to briefly describe the above algorithms and present their application to wireless communications optimization problems found in the literature. This chapter also presents results from different cases using PSO, ABC, ACO and DE. These include the cell to switch assignment problem in cellular networks using PSO algorithms, peak to average power ratio (PAPR) reduction of OFDM signals with the partial transmit sequences (PTS) approach using ABC and ACO algorithms [7, 11], and dual-band microwave filter design for wireless communications using SADE.

This chapter is subdivided into four sections. Section 2 presents the different evolutionary algorithms. Section 3 reviews the related work in wireless communications problems from the literature. Section 4 describes the design cases and presents the numerical results. Finally section 5 contains the discussion about the advantages of using a EA-based approach and the conclusions.

2. METHODS

A population (or swarm) in PSO, ABC, ACO and DE consists of *NP*vectors (or particles) $\bar{x}_{G,i}, i=1, 2, \ldots\ldots, NP,$, where *G*

is the generation number. The population is initialized randomly from a uniform distribution. Each *D*-dimensional vector represents a possible solution, which is expressed as:

$$\bar{x}_{G,i} = (x_{G,1i}, x_{G,2i}, \ldots x_{G,ji}, \ldots\ldots, x_{G,Di}) \tag{1}$$

The population is initialized as follows:

$$x_{0,ji} = rand_{j[0,1)}\left(x_{j,U} - x_{j,L}\right) + x_{j,L} \quad j = 1,2,\ldots\ldots,D \tag{2}$$

where *xj,L* and *xj,U* are *D*-dimensional vectors of the lower and upper bounds respectively and *randj*[0,1) is a uniformly distributed random number within [0,1). The stopping criterion for PSO, ABC and DE is usually the generation number or the number of objective-function evaluations.

2.1. Particle Swarm Optimization (PSO)

In PSO, the particles move in the search space, where each particle position is updated by two optimum values. The first one is the best solution (fitness) that has been achieved so far. This value is called *pbest*. The other one is the global best value obtained so far by any particle in the swarm. This best value is called *gbest*. After finding the *pbest* and *gbest*, the velocity update rule is an important factor in a PSO algorithm. The most commonly used algorithm defines that the velocity of each particle for every problem dimension is updated with the following equation:

$$u_{G+1,ni} = wu_{G,ni} + c_1 rand_{1(0,1)}(pbest_{G+1,ni} - x_{G,ni}) + c_2 rand_{2(0,1)}(gbest_{G+1,ni} - x_{G,ni}) \quad (3)$$

where $u_{G+1,ni}$

is the ith particle velocity in the nth dimension, $G+1$ denotes the current iteration and G the previous, xG,ni is the particle position in the nth dimension, $rand1(0,1), rand2(0,1)$ are uniformly distributed random numbers in (0,1), w is a parameter known as the inertia weight, and $c1$ and $c2$ are the learning factors.

The parameter w(inertia weight) is a constant between 0 and 1. This parameter represents the particle's fly without any external influence. The higher the value of w, or the closer it is to one, the more the particle stays unaffected from pbest and gbest. The inertia weight controls the impact of the previous velocity: a large inertia weight favors exploration, while a small inertia weight favors exploitation. The parameter $c1$ represents the influence of the particle memory on its best position, while the parameter $c2$ represents the influence of the swarm best position. Therefore, in the Inertia Weight PSO (IWPSO) algorithm the parameters to be determined are: the swarm size (or population size), usually 100 or less, the cognitive learning factor $c1$ and the social learning factor $c2$ (usually both are set to equal to 2.0), the inertia weight w , and the maximum number of iterations. It is common practice to linearly decrease the inertia weight starting from 0.9 or 0.95 to 0.4.

Clerc [8] suggested the use of a different velocity update rule, which introduced a parameter K

called constriction factor. The role of the constriction factor is to ensure convergence when all the particles have stopped their movement. The velocity update rule is then given by:

$$u_{G+1,ni} = K\left[u_{G,ni} + c_1 rand_{1(0,1)}(pbest_{G+1,ni} - x_{G,ni}) + c_2 rand_{2(0,1)}(gbest_{G+1,ni} - x_{G,ni})\right] \quad (4)$$

$$K = \frac{2}{\left|2 - \varphi - \sqrt{\varphi^2 - 4\varphi}\right|} \quad (5)$$

where $\varphi = c1 + c2$ and $\varphi > 4$ .This PSO algorithm variant is known as Constriction Factor PSO (CFPSO).

2.2. Barebones PSO

Kennedy [20] proposed a new PSO approach, the BB PSO, where the standard PSO velocity equation is replaced with samples from a normal distribution. In this method, the position update rule for the *i*th particle in the nth dimension becomes

$$x_{G+1,ni} = N\left(\frac{pbest_{G+1,ni} + gbest_{G+1,ni}}{2}, \left|pbest_{G+1,ni} - gbest_{G+1,ni}\right|\right) \tag{6}$$

N(,) denotes the normal distribution. The method allows particles with *pbest* significant different from *gbest* to make large step sizes towards it. When *pbest* is close to *gbest*, step size decreases and limits exploration in favor of exploitation.

In [20], a variation of BB PSO, the BBExp PSO, was also proposed. In this method, approximately half of the time velocity is based on samples from a normal distribution; for the rest of the time, velocity is derived from the particle's personal best position. The position update rule, (6), is modified into

$$x_{G+1,ni} = \begin{cases} N\left(\frac{pbest_{G+1,ni} + gbest_{G+1,ni}}{2}, \left|pbest_{G+1,ni} - gbest_{G+1,ni}\right|\right), & U(0,1) > 0.5 \\ pbest_{G+1,ni}, & \text{otherwise} \end{cases} \tag{7}$$

where *U*(,) denotes the uniform distribution. In BBExp PSO, position updates equal $pbest_n$ for half of the time resulting in the improved exploitation of $pbest_n$ compared to the BB PSO. One may notice that the barebones PSO algorithms do not require parameter tuning. More details can be found in [20, 21].

2.3. Artificial bee colony optimization

The ABC algorithm models and simulates the honey bee behavior in food foraging. In ABC algorithm, a potential solution to the optimization problem is represented by the position of a food source while the nectar amount of a food source corresponds to the quality (objective function fitness) of the associated solution. In order to find the best solution the algorithm defines three classes of bees: employed bees, onlooker bees and scout bees. The employed bee searches for the food sources, the onlooker bee makes a decision to choose the food sources by sharing the information of employed bee, and the scout bee is used to determine a new food source if a food source is abandoned by the employed bee and onlooker bee. For each food source exists only one employed bee (i.e. the number of the employed bees is equal to the number of solutions). The employed bees search for new neighbor food source near of their hive. A new position of the $\bar{x}_i=(x_{i,1},..,x_{i,j},..,x_{i,D})$ solution, where D is the problem dimension, is generated using

$$u_{i,j} = x_{i,j} + \varphi_{i,j}\left(x_{i,j} - x_{k,j}\right) \tag{8}$$

where $k \in \{1, 2, .., SN\}$, $k \neq i$, $j \in \{1, 2, .., D\}$ are randomly chosen indices, where SN is the number of food sources, and $\varphi_{i,j}$

is a uniformly distributed random number within [-1,1]. ABC uses a greedy selection operator, which for minimization problems is defined by

$$\bar{x}_i' = \begin{cases} \bar{u}_i, & \text{if } f(\bar{u}_i) < f(\bar{x}_i) \\ \bar{x}_i, & \text{otherwise} \end{cases} \tag{9}$$

where $\bar{x}_i'$ is the new position of the food source.

An onlooker bee chooses a food source depending on the probability value associated with that food source, p_i, given by

$$p_i = \frac{fit_i}{\sum_{m=1}^{SN} fit_m} \tag{10}$$

where fit_i

is the fitness value of the ith solution which is proportional to the nectar amount of the food source in the ith position. When a food source (solution) cannot be improved anymore then the scout bee helps the colony to randomly generate create new solutions

$$x_{i,j} = rand_{j(0,1)}\left(x_{j,U} - x_{j,L}\right) + x_{j,L} \quad j = 1,2,\ldots,D \tag{11}$$

where xj,L and xj,U are the lower and upper bounds of the jth dimension respectively and $randj(0,1)$ is a uniformly distributed random number within (0,1).

2.4. Ant colony optimization

Ant colony optimization (ACO) [6, 12, 22] is a meta-heuristic inspired by the ants' foraging behavior. At the core of this behavior is the indirect communication between the ants by means of chemical pheromone trails, which enables them to find short paths between their nest and food sources. Ants can sense pheromone. When they decide path to follow a path, they tend to choose the ones with strong pheromone intensities way back to the nest or to the food source. Therefore, shorter paths would accumulate more pheromone than longer ones. This feature of real ant colonies is exploited in ACO algorithms in order to solve combinatorial optimization problems considered to be NP-Hard.

2.5. Differential Evolution (DE)

The initial population evolves in each generation with the use of three operators: mutation, crossover and selection. Depending on the form of these operators several DE variants or strategies exist in the literature [14, 23]. The choice of the best DE strategy depends on problem type [24]. In SaDE the following four strategies are used for trial vector generation. These include *DE/rand/1bin*, *DE/rand-to-best/2/bin*, *DE/rand/2/bin*, and *DE/current-to-rand/1* [25]. In these strategies a mutant vector $v^{-}G+1,I$ for each target vector $x^{-}G,I$ is computed by:

$$DE / rand / 1 / bin$$

$$\overline{v}_{G+1,i} = \overline{x}_{G,r_1} + F(\overline{x}_{G,r_2} - \overline{x}_{G,r_3}), \quad r_1 \neq r_2 \neq r_3$$

$$DE / rand - to - best / 2 / bin$$

$$\overline{v}_{G+1,i} = \overline{x}_{G,i} + F(\overline{x}_{G,best} - \overline{x}_{G,i}) + F(\overline{x}_{G,r_1} - \overline{x}_{G,r_2}) + F(\overline{x}_{G,r_3} - \overline{x}_{G,r_4}), \quad r_1 \neq r_2 \neq r_3 \neq r_4$$

$$DE / rand / 2 / bin$$

$$\overline{v}_{G+1,i} = \overline{x}_{G,r_1} + F(\overline{x}_{G,r_2} - \overline{x}_{G,r_3}) + F(\overline{x}_{G,r_4} - \overline{x}_{G,r_5}), \quad r_1 \neq r_2 \neq r_3 \neq r_4 \neq r_5$$

$$DE / current - to - rand / 1 / bin$$

$$\overline{v}_{G+1,i} = \overline{x}_{G,i} + K(\overline{x}_{G,r_1} - \overline{x}_{G,i}) + F(\overline{x}_{G,r_2} - \overline{x}_{G,r_3}), \quad r_1 \neq r_2 \neq r_3 \tag{12}$$

where $r1, r2, r3, r4, r5$ are randomly chosen indices from the population, which are different from index i, F is a mutation control parameter, K a coefficient responsible for the level of recombination that occurs between $\bar{x}_{G,i}$ and $\bar{x}_{G,r1}$. After mutation, the crossover operator is applied to generate a trial vector $\bar{u}_{G+1,i} = (u_{G+1,1i}, u_{G+1,2i}, \ldots u_{G+1,ji}, \ldots, u_{G+1,Di})$ whose coordinates are given by:

$$u_{G+1,ji} = \begin{cases} v_{G+1,ji}, & \text{if } rand_{j[0,1)} \leq CR \text{ or } j=rn(i) \\ x_{G+1,ji}, & \text{if } rand_{j[0,1)} > CR \text{ and } j \neq rn(i) \end{cases} \tag{13}$$

where $j=1,2,\ldots\ldots,D$, $randj[0,1)$ is a number from a uniform random distribution from the interval $[0,1)$, rn(i) a randomly chosen index from $(1,2,\ldots\ldots,D)$, and CR the crossover constant from the interval $[0,1]$. DE uses a greedy selection operator, which for minimization problems is defined by:

$$\bar{x}_{G+1,i} = \begin{cases} \bar{u}_{G+1,i}, & \text{if } f(\bar{u}_{G+1,i}) < f(\bar{x}_{G,i}) \\ \bar{x}_{G,i}, & \text{otherwise} \end{cases} \tag{14}$$

where $f(\bar{u}_{G+1,i})$, $f(\bar{x}_{G,i})$ are the fitness values of the trial and the old vector respectively. Therefore, the newly found trial vector $\bar{u}_{G+1,i}$ replaces the old vector $\bar{x}_{G,I}$ only when it produces a lower objective-function value than the old one. Otherwise, the old vector remains in the next generation. The stopping criterion for the DE is usually the generation number or the number of objective-function evaluations.

2.6. Self-Adaptive DE (SADE)

In the SaDE algorithm both the trial vector generation strategies and the control parameters are self-adapted according to previous learning experiences. SaDE maintains a strategy candidate pool, consisting of the four strategies given in (2). Each strategy is assigned a certain probability. The sum of all probabilities is equal to one. These probabilities are initialized with a value of 0.25 and gradually adapted during evolution. The probability of applying the *m-th* strategy is $pm, m=1,2,\ldots\ldots,M$, where M is the total number of strategies. At generation G, the number of successful trial vectors generated by the *m-th* strategy is denoted as nsm,G, while the number of trial vectors that fail to replace the old vectors in the next generation is *nfm,G*. An additional parameter called the learning period (LP) is introduced in [17]. This corresponds to the number of the previous generations that store the success and fail statistics. After LP generations, the probabilities of selecting different strategies are updated according to:

$$p_{m,G} = \frac{S_{m,G}}{\sum_{m=1}^{M} S_{m,G}}$$

where

$$S_{m,G} = \frac{\sum_{g=G-LP}^{G-1} ns_{m,g}}{\sum_{g=G-LP}^{G-1} ns_{m,g} + \sum_{g=G-LP}^{G-1} nf_{m,g}} + \varepsilon \qquad (15)$$

where *Sm,G* is the success rate of the trial vectors generated by the *m-th* strategy within the previous LP generation and ε is a constant set equal to 0.01 to avoid possible null success rates. Therefore, according to (5) strategies with high success rates have higher probability to be applied at the current generation.

The control parameters are self-adapted in the following way. The mutation control parameter *F* is approximated by a normal distribution with mean value 0.5 and standard deviation 0.3, that is *N*(0.5,0.3). The parameter *K* is a random number in the interval [0, 1] generated by a uniform distribution. The crossover rate control parameter *CR* used by the *m-th* strategy is also approximated by a normal distribution with mean value *CRm* and standard deviation 0.1, that is *N(CRm*,0.1). The initial value of *CRm* is 0.5 for all strategies. The values of crossover rates that have successfully generated trial vectors in the previous LP generations are stored in a crossover rate memory for each strategy *CRmmemory* that is an array of size LP. At each generation, the median value stored in memory for the *m-th* strategy *CRmmedian* is calculated and the *CR* values generated are given by a normal distribution with mean value *CRmmedian* and standard deviation 0.1. That way the crossover values are evolved at each generation to follow the successful values found. The authors in [17] suggest a value between 20 and 60 for the parameter LP. The sensitivity analysis performed in [17] for the LP parameter showed it had no significant impact on SaDE performance. More details about the SaDE algorithm can be found in [17].

3. RELATED WORK

This section presents a brief literature review of applications of evolutionary algorithms and their variants to wireless communications problems.

Genetic algorithms (GAs) are among the widely used optimization techniques for addressing design problems in wireless communications. In [26] a Smith prediction filter is proposed for power control design of direct-sequence code-division multiple-access cellular mobile radio systems. A fixed-order robust $H\infty$ loop filter is developed using a genetic algorithm to minimize the worst-case variance of the received SINR from the minimax perspective. The authors in [27] present an antenna selection method for multiple-input multiple-output wireless systems based on a GA that seeks the best subset of antenna elements. The problem of receive antenna selection and symbol detection for multiple-input, multiple-output (MIMO) systems is solved in [28] by applying a genetic algorithm (GA) variant. The paper in [29] addresses the problem of joint transmit/receive antenna selection for MIMO systems using a real-valued genetic algorithm (RVGA). The optimization objective is to improve the channel capacity of multiple-input/multiple-output (MIMO) systems. The study in [30] presents a pattern discovery algorithm for multi-streams mining in wireless sensor networks. This algorithm adapts genetic operators with Elitism Strategy.The paper in [31] studies joint multiuser linear precoding design in the forward link of fixed multibeam satellite systems. The authors use a generic optimization framework for linear precoding design to handle any objective functions of data rate with general linear and nonlinear power constraints. In [32] an energy-efficient genetic algorithm mechanism is presented to resolve quality of service (QoS) multicast routing problem: The proposed genetic algorithm depends on bounded end-to-end delay and minimum energy cost of the multicast tree.

SI algorithms are among the most commonly used algorithms for solving problems in wireless communications. PSO and several PSO variants have been used in the literature to solve different problems. In [33] optimal power scheduling for distributed detection in a Gaussian sensor network is addressed for both independent and correlated observations. A PSO based technique is developed to find the optimal power allocation for arbitrary correlations. The authors in [34] apply PSO to solve the constrained nonlinear optimisation problem for the minimum bit error rate (MBER) multiuser transmitter (MUT). The proposed PSO aided symbol-specific MBER-MUT and average MBER-MUT schemes provide improved performance in comparison to the conventional minimum mean-square error MUT scheme. Several issues in Wireless Sensor Networks (WSNs) can be formulated as multidimensional optimization problems, and addressed using bio-inspired algorithms. In [35] the authors present a brief survey of how PSO is used to address these issues. The authors in [36] propose a new approach to estimate the location of a sensor in a wireless sensor network based on a new PSO algorithm with a log-barrier approach. The paper in [37] presents a predistorter based on a cluster-based implementation particle swarm optimization technique with embedded model-size estimation capability and validates the proposed technique on a Doherty power amplifier prototype.

The ABC algorithm and its variants have been successfully applied to several optimization problems in wireless communications. Among others these include issues in WSN [38-41], WiMax network planning [42] and channel assignment [43].

The ACO algorithm has also been applied to several combinatorial problems in wireless systems which include problems in mobile ad hoc networks [44-46], problems in WSN [47-51], cognitive radio [52], resource allocation in multiuser OFDM systems [53] and MIMO problems [54, 55].

DE variants have also been applied to variety of optimization problems like multi-user detection in multi-carrier CDMA [56], WSNs issues [57, 58], urban area path loss prediction [59], spectrum sharing [60], optimization of interleave-division multiple-access communication systems [61]. Other papers use a number of different optimization algorithms and compare results. For example in [62] spectrum allocation methods for cognitive radio based on GA, quantum genetic algorithm (QGA), and PSO, are proposed.

4. RESULTS AND DISCUSSIONS

In this section we present numerical results from different optimization problems in wireless communications using different algorithms.

4.1. Cell to switch assignment in cellular networks using barebones PSO

Cell assignment is an important issue in the area of resource management in cellular networks. The problem is an NP-hard one and requires efficient search techniques for its solution in real-time. We briefly present an example case of solving this problem using the barebones PSO [9]. The effective assignment of cells to switches in order to minimize the cost of network deployment is a challenging issue in cellular networking.

The cell-to-switch assignment (CSA) problem consists of optimally assigning cells to network switches while respecting certain constraints such as the call volume of each cell and the switches capacity [63]. The objective of the optimization is the reduction of implementation and operational costs. Usually, the cost function considers the cost of linking cells to switches (cabling cost) and the cost of handoff between different cells (handoff cost). The problem is an NP-hard one with exponential complexity and cannot be solved analytically in real size networks. Other evolutionary techniques like tabu search [64], ACO

[65] and GAs [66] have been used in the literature for solving this problem. The problem formulation is given below.

We consider n unique and distinct cells in a given service area and m switches with known location and traffic parameters. The objective is the optimum assignment of cells to switches in order to minimize the total cost that comprises the handoff and cabling costs.

In single-homing CSA, each cell belongs to one cluster and it is assigned to one switch at a time. In this case, the objective function to be minimized is [63]:

$$\sum_{i=1}^{n}\sum_{j=1}^{m} c_{ik}x_{ik} + \sum_{i=1}^{n}\sum_{\substack{j=1\\ j\neq i}}^{n} h_{ij}\left(1-y_{ij}\right), \quad k=1,\ldots,m \tag{16}$$

where c_{ik} is the cabling cost per time unit between cell i and switch k, x_{ik} is a parameter that takes the value one when cell i is assigned to switch k (otherwise, x_{ik}=0) and h_{ij} is the cost per time unit for the handoffs that occur between cells i and j. The first term in (16) gives the total cabling cost and the second one is the total handoff cost per time unit among cells. Therefore, y_{ij} is defined as

$$y_{ij} = \sum_{k=1}^{m} x_{ik}x_{jk}, \quad i,j=1,\ldots,n \tag{17}$$

where y_{ij} is one when cells i and j are connected to the same switch, otherwise it is zero. The product $x_{ik}x_{jk}$ in (17) defines the variable

$$z_{ijk}=x_{ik}x_{jk} \quad i,j=1,\ldots,n \text{ and } k=1,\ldots,m \tag{18}$$

that is zero unless cells i and j are connected to switch k. In this case, it is one. Cell assignment is subject to further constraints. The call handling capacity of each switch should not be violated at any time, i.e.

$$\sum_{k=1}^{n} \lambda_i x_{ik} \le M_k, \quad i=1,\ldots,n \tag{19}$$

where λi is the number of calls that cell i handles per unit time and M_k is the call handling capacity of switch k. Also, each cell is assigned only to one switch, i.e.

$$\sum_{k=1}^{m} x_{ik} = 1, \quad i=1,\ldots,n \tag{20}$$

The optimization problem defined by (16) and subject to (17)-(20), can be converted [63] to an integer programming one by replacing (18) with the

$$\begin{array}{ll} 0 \le z_{ijk} \le x_{jk}, x_{ik} & \\ z_{ijk} \ge x_{ik} + x_{jk} - 1, & i,j = 1,\ldots,n \text{ and } k = 1,\ldots,m \end{array} \tag{21}$$

We compare BB and BBExp PSO with ACO [65] and binary PSO (BPSO) [67]. We run 100 independent trials for each algorithm. The statistical results are presented and compared In the proposed barebones PSO variants, the only parameter we set was the swarm size. In the examples presented here, this was set to 5 particles. The ACO parameters were the same as in [65]. For the BPSO, we set the learning factors c_1 and c_2 equal to two. Systems with varied number of cells and switches that range from 15 to 200 and from 2 to 7, respectively, were considered.

The percentage of successfully obtained solutions as a function of the number of cells and switches indicates the solution accuracy of the algorithms. Figure 1 shows the results of the application of BB PSO, BBExp PSO, BPSO and ACO in a single-homing system for swarm size equal to 5 particles. In the first case, BB PSO outperforms the other methods in systems with small complexity but as the complexity increases (n/m=150/6 and 200/7) BBExp gives the best results. As it was expected, solution accuracy decreases with system complexity. In general, BB PSO outperforms the other methods for small n and m. However, its performance degrades with system complexity; in this case, BBExp gives better results. In any case, at least one of BB and BBExp PSO is better than BPSO and ACO. We have also evaluated the different algorithms in terms of the computational time required for the derivation of the previous results. Figure 2 shows small differences in results between the four algorithms. In all cases, barebones PSO algorithms are slightly faster. BBExp outperforms BB as system complexity increases. The computational cost of the methods grows exponentially with n/m and increases with the swarm size. More details about this problem can be found in [9].

4.2. A PTS technique based on ACO and ABC for PAPR reduction of OFDM signals

A major drawback of OFDM signals is the high value of peak to average power ratio (PAPR). Partial transmit sequences (PTS) [68], is a popular PAPR reduction method with good PAPR reduction performance. However, PTS requires an exhaustive search in order to find the optimal phase factors. Thus, the search complexity is high. Several methods have been published in the literature for PAPR reduction using PTS with low search complexity [10, 69, 70]. The problem description is given below.

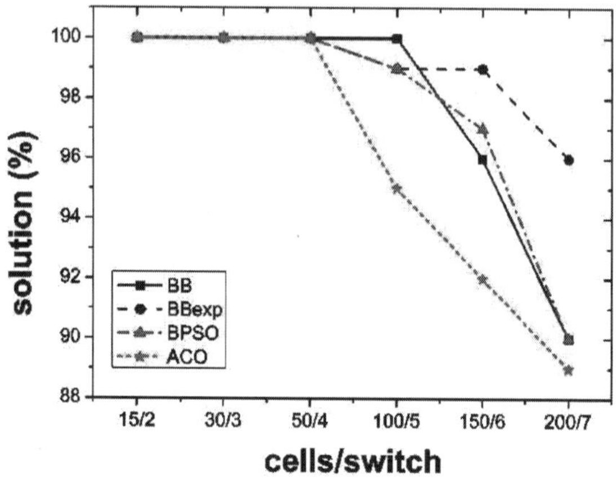

Figure 1. Successful solutions vs cells/switch

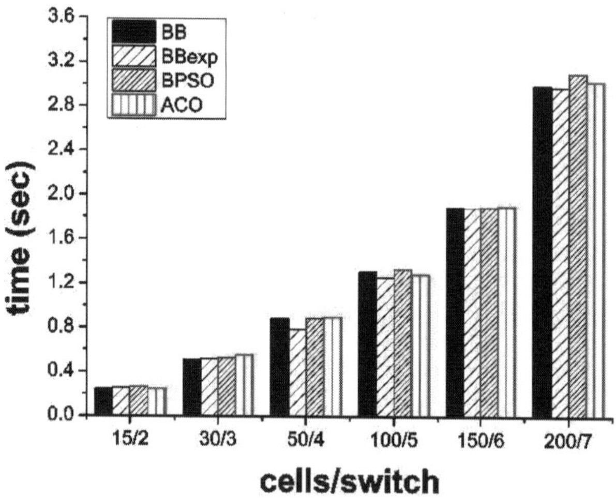

Figure 2. Computational time vs cells/switch chart

In an OFDM system, the high-rate data steam is split into N low-rate data streams that are simultaneously transmitted using N subcarriers. The discrete-time signal of such a system is given by

$$s_k = \frac{1}{\sqrt{N}} \sum_{n=0}^{N-1} S_n e^{\frac{j2\pi nk}{LN}} \quad k = 0,1,....,LN-1 \tag{22}$$

where L is the oversampling factor, **S**=[$S0$,$S1$,...,$SN-1$]T is the input signal block. Each symbol is modulated by either phase-shift keying (PSK) or quadrature amplitude modulation (QAM).

The PAPR of the signal in (22) is defined as the ratio of the maximum to average power and is expressed in dB as

$$PAPR(s) = 10\log_{10} \frac{\max\limits_{0 \le k \le LN-1} |s_k|^2}{E\left[|s_k|^2\right]} \tag{23}$$

where E[.]is the expected value operation. In the PTS approach the **S** input data OFDM block is partitioned into M disjointed subblocks represented by the vector **S**mm=1,2,...,M−1 and oversampled by inserting $(L-1)N$ zeros. Then the PTS process is expressed as

$$S = \sum_{m=1}^{M} S_m \tag{24}$$

Next, the subblocks are converted to time domain using LN point inverse fast fourier transform (IFFT). The representation of the OFDM block in time domain is expressed by

$$s = IFFT\left\{\sum_{m=1}^{M} S_m\right\} = \sum_{m=1}^{M} IFFT\{S_m\} = \sum_{m=1}^{M} s_m \tag{25}$$

The PTS objective is to produce a weighted combination of the M subblocks using $\mathbf{b}=[b1,b2,...,bM]^T$ complex phase factors to minimize PAPR. The transmitted signal in time domain after this combination is given by

$$s'(b) = \sum_{m=1}^{M} b_m s_m \tag{26}$$

In order to reduce the search complexity the phase factor possible values are limited to a finite set. The set of allowable phase factors is

$$A = \left\{ e^{\frac{j2\pi n}{W}} \middle| n = 0,1,...,W-1 \right\} \tag{27}$$

where W is the number of allowed phase factors. Therefore in case of M subblocks and W phase factors the total number of possible combinations is WM. In order to reduce the search complexity we usually set fixed one phase factor.

The optimization goal of the PTS scheme is to find the optimum phase combination for minimum PAPR. Thus, the objective function can be expressed as

Minimize

$$F(b) = 10\log_{10} \frac{\max\limits_{0 \le k \le LN-1} \left| s'(b) \right|^2}{E\left[\left| s'(b) \right|^2 \right]} \tag{28}$$

subject to

$$b \in \left\{ e^{j\phi_m} \right\}^M \text{ where } \phi_m \in \left\{ \frac{2\pi l}{W} \middle| l = 0,1,...,W-1 \right\} \tag{29}$$

We have evaluated objective function above using evolutionary algorithms and methods found in the literature. We have used two main measurement criteria namely the complementary cumulative distribution function (CCDF) and the computational complexity. In all our simulations, 10E5 random OFDM blocks are generated. The transmitted signal is oversampled by a factor L=4. We consider 16-QAM modulation with N=256 sub-carriers which are divided into M=16 random subblocks. The phase factors for W=2 are selected. We consider the first phase factor to be fixed so the total number of unknown phase factors is M-1.

The control parameters in all simulations are given below. In the PSO algorithm $c1$ and $c2$ are set equal to 2.05 while the inertia weight is linearly decreased starting from 0.9 to 0.4. For ACO the initial pheromone value $\tau 0$ is set to 1.0e-6, the pheromonone update constant Q is set to 20, the exploration constant $q0$ is set to 1, the global pheromone decay rate ρg is 0.9, the local pheromone decay rate ρl is 0.5, the pheromone sensitivity α is 1, and the visibility sensitivity is β is 5.

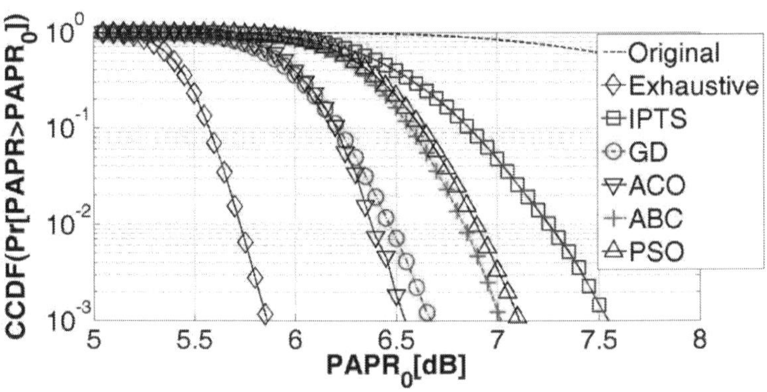

Figure 3.PARP reduction performance comparison of the BBO-PTS algorithms with other PTS schemes for NP=30, G=30.

Figure 3 presents the comparison between the CCDF by different PTS reduction techniques. For Pr(*PAPR*>*PAPR*0)=10^{-3} the PAPR of the original OFDM transmitted signal is 10.84dB. For all evolutionary algorithms, the population size NP is set to 30 and the maximum number of generations G is set to 30. Thus, the computational complexity of this case is $NP\times G$=900. The computational complexity of the exhaustive search is $WM-1$=32768 while the PAPR for this case is 5.86dB. The PAPR by the iterative flipping algorithm for PTS (IPTS) [69] is 7.55dB with search complexity (M-1)W=30. The PAPR by the gradient descent method (GD) [71] with search complexity *CrM*−1*WrI*=*C*215223=1260 is 6.96dB. The PAPR by ABC [10], PSO [72], and ACO [6], is 7.01dB, 7.13dB, and 6.52dB, respectively. Table 1 holds the comparison of the search complexity among the different methods for

CCDF=10^{-3} NP=30, and G=30. It is obvious that ACO presents the better performance among the other methods with the same search complexity.

Table 1. Comparison of computational complexity for CCDF=1e-3 among different PTS Schemes

Method	Computational Complexity	PAPR (dB)
Original	0	10.84
Exhaustive	$W^{M-1}=32768$	5.86
IPTS	$(M-1)W=30$	7.55
GD	$C^{r}_{M-1}W^{r}I=C^{2}_{15}2^{2}3=1260$	6.96
PSO	$NP\times G=900$	7.13
ABC	$NP\times G=900$	7.01
ACO	$NP\times G=900$	6.52

4.3. Dual-band microwave filter design using SADE

Microwave filters are among the important components of a modern wireless communication system. Several papers exist in the literature that address the filter design problem [73]. Open Loop Ring Resonator (OLRR) filters, which consist of two uniform microstrip lines and pairs of open loops between them, are widely used as the building block in several multiband bandpass filter design cases [74]. In [74], two pairs of folded OLRRs operating at two passbands are proposed to produce dual-band response.

A dual band OLRR filter is shown in Figure 4. The frequency response of such a filter depends on the filter dimensions and spacings between microstrip lines [74, 75]. The design parameters for this case are the ones shown in Figure 4, $(W1,W2,L1,L2,L3,L4,L5,S1,S2,S3,G)$, all expressed in mm.

Such a filter design problem can be defined by the minimization of $|S11|$ in the passband frequency range. This design problem is therefore defined by the minimization of the objective function:

$$\mathrm{F}(\overline{x}) = 20\log\left\{\max\left|S_{11}(\overline{x},f)\right|,\ f\in S_{p}\right\} \tag{30}$$

where $\overline{x}$ is the vector of filter geometry parameters and *Sp* is the set of distinct frequencies in the desired passband frequency ranges.

The filter is designed for operation in two WiMax (IEEE 802.16) frequency bands. These are the 3.5GHz and the 5.8GHz frequency bands. For this case, we set *Sp*={3.55,3.6,5.75,5.8} . Figure 5 shows the simulated frequency response of this design. The simulated current distribution for the 3.6GHz and 5.8GHz frequencies is presented in Figure 6, where the resonating ring in each case is clearly seen. In the first passband between 3.508 and 3.809 GHz, the filter has a return loss less than 10dB and insertion loss greater than 0.5dB. In the second passband between 5.744 and 6.121 GHz the results also show a return loss less than 10dB and insertion loss greater than 0.5dB. The rejection band (between 4.236 and 5.367 GHz) has an insertion loss less than 20dB. In the first passband the return loss is less than 29dB at both 3.533 GHz and 3.759 GHz. In the second passband the return loss is less than 22dB at 5.794GHz and less than 28dB at 6.07GHz.

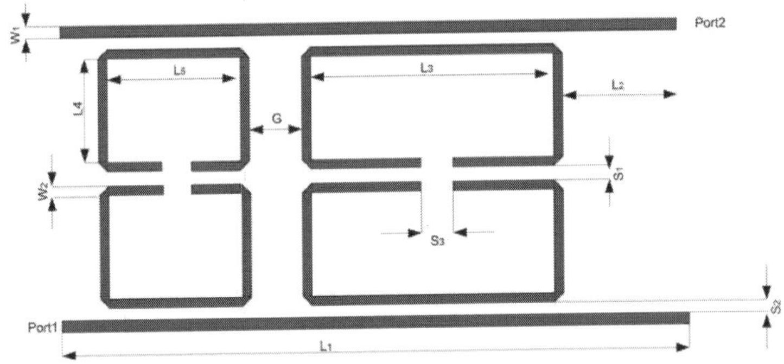

Figure 4. Dual-band filter geometry

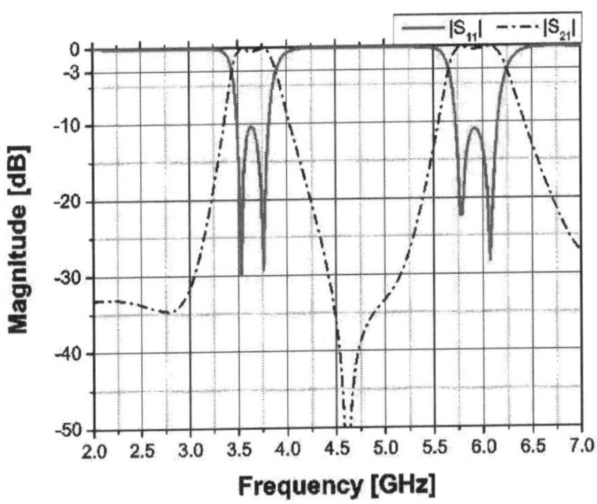

Figure 5. Simulated frequency response of the dual-band filter.

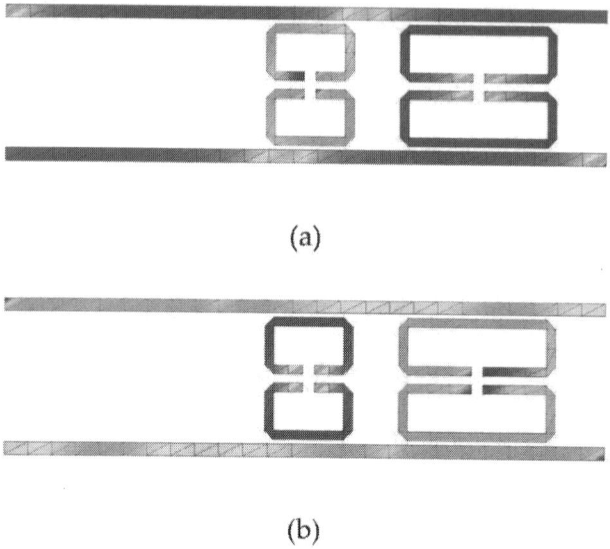

Figure 6. Current distribution simulations for dual-band filter at (a) 3.5GHz and (b) 5.8GHz

5. CONCLUSION

A brief survey of different evolutionary algorithms and their application to different problems in wireless communications has been presented. It must be pointed out that several evolutionary algorithms exist in the literature. GAs and SI algorithms are among those most commonly used. In order to select, the best algorithm for every problem one has to consider the problem characteristics. Another key issue is the selection of the algorithm control parameters, which is also in most cases problem-dependent. One may also use at first the control parameters for these algorithms that commonly perform well regardless of the characteristics of the problem to be solved. The example of the CSA problem in cellular networks showed the better performance of BB and BBExp compared to BPSO and ACO in terms of successfully obtained solutions and execution time. In the PTS optimization problem ACO outperformed ABC and PSO.

The selection of the SADE technique for microwave filter design has lead to a successful filter design which exhibits low loss in the passbands and high isolation between the passbands. The DE algorithms are also robust optimizers. In classical DE algorithms, the selection of the appropriate strategy for trial vector generation and control parameters requires additional computational time using a trial-and-error search procedure. Therefore, it is not always an easy task to fine-tune the control parameters and strategy given also that commonly the appropriate control parameters and strategy selection are problem dependent. The SaDE advantage though, is the fact that no additional time for solving a given problem is required. SaDE requires only the adjustment of two parameters: the population size and the number of iterations.

The practical examples subject to several constraints presented in this chapter show the applicability and the efficiency of using such algorithms.

REFERENCES

1. Goldberg D. E. Genetic algorithms in search, optimization and machine learning. New York:Addison Wesley;1989.
2. Holland J. H. Adaptation in natural and artificial systems. Ann Arbor: The University of Michigan Press;1975.
3. Fogel D. B. Evolutionary computation: Toward a new philosophy of machine intelligence. IEEE Press, Piscataway, NJ;1995.
4. Beyer H. G. and Schwefel H. P. Evolution strategies: A comprehensive introduction. Natural Computing 2002;1 (1) 3-52.
5. Kennedy J. and Eberhart R., "Particle swarm optimization", in IEEE International Conference on Neural Networks, pp. 1942-1948, Piscataway, NJ, 1995.
6. Dorigo M. and Stutzle T. Ant colony optimization. Cambridge, MA:The MIT Press;2004.
7. Karaboga D. and Basturk B. A powerful and efficient algorithm for numerical function optimization: Artificial bee colony (abc) algorithm. Journal of Global Optimization 2007;39 (3) 459-471.
8. Clerc M., "The swarm and the queen: Towards a deterministic and adaptive particle swarm optimization", in Proceedings of the 1999 Congress on Evolutionary Computation, 1999. CEC 99 pp. 1951-1957, Washington, DC, 1999.
9. Goudos S. K., Baltzis K. B., Bachtsevanidis C. and Sahalos J. N. Cell-to-switch assignment in cellular networks using barebones particle swarm optimization. IEICE Electron. Express 2010;7 (4) 254-260.
10. Wang Y., Chen W. and Tellambura C. A papr reduction method based on artificial bee colony algorithm for ofdm signals. IEEE Transactions on Wireless Communications 2010;9 (10) 2994-2999.
11. Zhu G. and Kwong S. Gbest-guided artificial bee colony algorithm for numerical function optimization. Applied Mathematics and Computation 2010;217 (7) 3166-3173.

12. Dorigo M., Maniezzo V. and Colorni A. Ant system: Optimization by a colony of cooperating agents. IEEE Transactions on Systems, Man, and Cybernetics, Part B: Cybernetics 1996;26 (1) 29-41.

13. Storn R. and Price K., "Differential evolution—a simple and efficient adaptive scheme for global optimization over continuous spaces", in Tech. Rep. TR-95-012, Berkeley, CA, 1995.

14. Storn R. and Price K. Differential evolution-a simple and efficient heuristic for global optimization over continuous spaces. Journal of Global Optimization 1997;11 (4) 341-359.

15. Qin A. K. and Suganthan P. N., "Self-adaptive differential evolution algorithm for numerical optimization", in The 2005 IEEE Congress on Evolutionary Computation, 2005., ed. P. N. Suganthan, pp. 1785-1791, 2005.

16. Huang V. L., Qin A. K. and Suganthan P. N., "Self-adaptive differential evolution algorithm for constrained real-parameter optimization", in IEEE Congress on Evolutionary Computation, 2006. CEC 2006., ed. A. K. Qin, pp. 17-24, 2006.

17. Qin A. K., Huang V. L. and Suganthan P. N. Differential evolution algorithm with strategy adaptation for global numerical optimization. IEEE Transactions on Evolutionary Computation 2009;13 (2) 398-417.

18. Goudos S. K., Zaharis Z. D. and Yioultsis T. V. Application of a differential evolution algorithm with strategy adaptation to the design of multi-band microwave filters for wireless communications. Progress in Electromagnetics Research 2010;109 (123-137.

19. Goudos S. K., Siakavara K., Samaras T., Vafiadis E. E. and Sahalos J. N. Sparse linear array synthesis with multiple constraints using differential evolution with strategy adaptation. IEEE Antennas and Wireless Propagation Letters 2011;10 (670-673.

20. Kennedy J. Bare bones particle swarms. Proceedings of the IEEE Swarm Intelligence Symposium 2003 (SIS 2003) 2003;80-87.

21. Pan F., Hu X., Eberhart R. and Chen Y., "An analysis of bare bones particle swarm", 2008.

22. Dorigo M. and Gambardella L. M. Ant colonies for the travelling salesman problem. BioSystems 1997;43 (2) 73-81.

23. Storn R. Differential evolution research-trends and open questions. Studies in Computational Intelligence 2008;143 (1-31.

24. Mezura-Montes E., Velazquez-Reyes J. and Coello Coello C. A., "A comparative study of differential evolution variants for global

optimization", in GECCO 2006-Genetic and Evolutionary Computation Conference, pp. 485-492, Seattle, WA, 2006.

25. Iorio A. W. and Li X. Solving rotated multi-objective optimization problems using differential evolution. Lecture Notes in Artificial Intelligence (Subseries of Lecture Notes in Computer Science) 2004;3339 (861-872.

26. Lee B. K., Chen H. W. and Chen B. S. Power control of cellular radio systems via robust smith prediction filter. IEEE Transactions on Wireless Communications 2004;3 (5) 1822-1831.

27. Karamalis P. D., Skentos N. D. and Kanatas A. G. Selecting array configurations for mimo systems: An evolutionary computation approach. IEEE Transactions on Wireless Communications 2004;3 (6) 1994-1998.

28. Lu H. Y. and Fang W. H. Joint receive antenna selection and symbol detection for mimo systems: A heterogeneous genetic approach. IEEE Communications Letters 2009;13 (2) 97-99.

29. Lain J. K. Joint transmit/receive antenna selection for mimo systems: A real-valued genetic approach. IEEE Communications Letters 2011;15 (1) 58-60.

30. Cheng W., Shi H., Yin X. and Li D. An elitism strategy based genetic algorithm for streaming pattern discovery in wireless sensor networks. IEEE Communications Letters 2011;15 (4) 419-421.

31. Zheng G., Chatzinotas S. and Ottersten B. Generic optimization of linear precoding in multibeam satellite systems. IEEE Transactions on Wireless Communications 2012;11 (6) 2308-2320.

32. Lu T. and Zhu J. Genetic algorithm for energy-efficient qos multicast routing. IEEE Communications Letters 2013;17 (1) 31-34.

33. Wimalajeewa T. and Jayaweera S. K. Optimal power scheduling for correlated data fusion in wireless sensor networks via constrained pso. IEEE Transactions on Wireless Communications 2008;7 (9) 3608-3618.

34. Yao W., Chen S., Tan S. and Hanzo L. Minimum bit error rate multiuser transmission designs using particle swarm optimisation. IEEE Transactions on Wireless Communications 2009;8 (10) 5012-5017.

35. Kulkarni R. V. and Venayagamoorthy G. K. Particle swarm optimization in wireless-sensor networks: A brief survey. IEEE Trans Syst Man Cybern Pt C Appl Rev 2011;41 (2) 262-267.

36. Nguyen H. A., Guo H. and Low K. S. Real-time estimation of sensor node's position using particle swarm optimization with log-barrier constraint. IEEE Trans. Instrum. Meas. 2011;60 (11) 3619-3628.

37. Abdelhafiz A. H., Hammi O., Zerguine A., Al-Awami A. T. and Ghannouchi F. M. A pso based memory polynomial predistorter with embedded dimension estimation. IEEE Transactions on Broadcasting 2013;59 (4) 665-673.

38. Li M., Xiong W. and Liang Q. Wireless sensor networks node localization algorithm based on improved abc algorithm. Chin. J. Sens. Actuators 2013;26 (2) 241-245.

39. Li M., Xiong W. and Liang Q., "An improved abc-based node localization algorithm for wireless sensor network", in 2012 8th International Conference on Wireless Communications, Networking and Mobile Computing, WiCOM 2012, Shanghai, 2012.

40. Hè P. Y. and Jiang M. Y., "Dynamic deployment of wireless sensor networks by an improved artificial bee colony algorithm", in 2013 International Conference on Sensors, Mechatronics and Automation, ICSMA 2013, pp. 862-866, Shenzhen, 2014.

41. Gu Y., Xu X., Du J., Hou R. and Qian H. Anycast routing protocol for wireless sensor networks based on artificial bee colony. Chin. J. Sens. Actuators 2013;26 (4) 564-569.

42. Berrocal-Plaza V., Vega-Rodríguez M. A., Gómez-Pulido J. A. and Sánchez-Pérez J. M., "Artificial bee colony algorithm applied to wimax network planning problem", in 2011 11th International Conference on Intelligent Systems Design and Applications, ISDA'11, pp. 504-509, Cordoba, 2011.

43. Liu J., Jia Z., Qin X., Chang C., Xu G. and Xia X., "The applications in channel assignment based on cooperative hybrid artificial bee colony algorithm", in 2011 International Conference on Electrical Engineering and Automation, EEA 2011, pp. 401-406, Beijing, 2012.

44. Deepalakshmi P. and Radhakrishnan S. An ant colony-based, receiver-initiated multicast mesh protocol for collaborative applications of mobile ad hoc networks. Eur Trans Telecommun 2014;25 (3) 354-369.

45. Balaji V. and Duraisamy V. Ant optimized link quality for ad hoc on demand distance vector. Wireless Pers Commun 2014;

46. Shirkande S. D. and Vatti R. A., "Aco based routing algorithms for ad-hoc network (wsn,manets):A survey", in 3rd International Conference on Communication Systems and Network Technologies, CSNT 2013, pp. 230-235, Gwalior, 2013.

47. Liu X. and He D. Ant colony optimization with greedy migration mechanism for node deployment in wireless sensor networks. J Network Comput Appl 2014;39 (1) 310-318.

48. Liu X. A transmission scheme for wireless sensor networks using ant colony optimization with unconventional characteristics. IEEE Communications Letters 2014;18 (7) 1214-1217.

49. Liu W. and Wang L. Ant colony optimization routing algorithm based on wsn. ICIC Express Lett Part B Appl. 2013;4 (3) 541-547.

50. Li Z. and Shi Q., "An qos algorithm based on aco for wireless sensor network", in 15th IEEE International Conference on High Performance Computing and Communications, HPCC 2013 and 11th IEEE/IFIP International Conference on Embedded and Ubiquitous Computing, EUC 2013, pp. 1671-1674, IEEE Computer Society, Zhangjiajie, Hunan, 2014.

51. Liu X. Sensor deployment of wireless sensor networks based on ant colony optimization with three classes of ant transitions. IEEE Communications Letters 2012;16 (10) 1604-1607.

52. He Q., Feng Z. and Zhang P. Reconfiguration decision making based on ant colony optimization in cognitive radio network. Wireless Pers Commun 2013;71 (2) 1247-1269.

53. Zhao Y., Xu X., Hao Z., Tao X. and Zhang P., "Resource allocation in multiuser ofdm system based on ant colony optimization", in IEEE Wireless Communications and Networking Conference 2010, WCNC 2010, Sydney, NSW, 2010.

54. Marinello J. C. and Abrao T., "Lattice reduction aided detector for dense mimo via ant colony optimization", in 2013 IEEE Wireless Communications and Networking Conference, WCNC 2013, pp. 2839-2844, Shanghai, 2013.

55. Lain J. K. and Chen J. Y. Near-mld mimo detection based on a modified ant colony optimization. IEEE Communications Letters 2010;14 (8) 722-724.

56. Das S., Mukherjee R., Kundu R. and Vasilakos T., "Multi-user detection in multi-carrier cdma wireless broadband system using a binary adaptive differential evolution algorithm", in 2013 15th Genetic and Evolutionary Computation Conference, GECCO 2013, pp. 1245-1252, Amsterdam, 2013.

57. Kundu S., Das S., Vasilakos A. V. and Biswas S. A modified differential evolution-based combined routing and sleep scheduling scheme for lifetime maximization of wireless sensor networks. Soft Computing 2014;

58. Yin X., Ling Z., Guan L. and Liang F. Minimum distance clustering algorithm based on an improved differential evolution. Int. J. Sens. Netw. 2014;15 (1) 1-10.

59. Wu C. Y., Yen H. S., Chiu C. C., Chiang J. S. and Chang C. W., "Urban area propagation path loss reduction by dynamic differential evolution algorithm", in 1st International Conference on Intelligent Green Building and Smart Grid, IGBSG 2014, IEEE Computer Society, Taipei, 2014.

60. Wang Z. and Zhang W. Spectrum sharing with limited channel feedback. IEEE Transactions on Wireless Communications 2013;12 (5) 2524-2532.

61. Li K., Wang X. and Ping L. Analysis and optimization of interleave-division multiple-access communication systems. IEEE Transactions on Wireless Communications 2007;6 (5) 1973-1982.

62. Zhao Z., Peng Z., Zheng S. and Shang J. Cognitive radio spectrum allocation using evolutionary algorithms. IEEE Transactions on Wireless Communications 2009;8 (9) 4421-4425.

63. Merchant A. and Sengupta B. Assignment of cells to switches in pcs networks. IEEE/ACM Transactions on Networking 1995;3 (5) 521-526.

64. Pierre S. and Houéto F. A tabu search approach for assigning cells to switches in cellular mobile networks. Computer Communications 2002;25 (5) 464-477.

65. Shyu S. J., Lin B. M. T. and Hsiao T. S. Ant colony optimization for the cell assignment problem in pcs networks. Computers and Operations Research 2006;33 (6) 1713-1740.

66. Salcedo-Sanz S. and Yao X. Assignment of cells to switches in a cellular mobile network using a hybrid hopfield network-genetic algorithm approach. Appl. Soft Comput. J. 2008;8 (1) 216-224.

67. Kennedy J. and Eberhart R. C., "Discrete binary version of the particle swarm algorithm", in Proceedings of the IEEE International Conference on Systems, Man and Cybernetics, pp. 4104-4108, 1997.

68. Müller S. H. and Huber J. B. Ofdm with reduced peak-to-average power ratio by optimum combination of partial transmit sequences. Electronics Letters 1997;33 (5) 368-369.

69. Cimini Jr L. J. and Sollenberger N. R. Peak-to-average power ratio reduction of an ofdm signal using partial transmit sequences. IEEE Communications Letters 2000;4 (3) 86-88.

70. Tellambura C. Improved phase factor computation for the par reduction of an ofdm signal using pts. IEEE Communications Letters 2001;5 (4) 135-137.

71. Han S. H. and Lee J. H. Papr reduction of ofdm signals using a reduced complexity pts technique. IEEE Signal Processing Letters 2004;11 (11) 887-890.

72. Wen J. H., Lee S. H., Huang Y. F. and Hung H. L. A suboptimal pts algorithm based on paticle swarm optimization for papr reduction in ofdm systems. EURASIP J. Wireless Commun. Networking 2008;2008 (14)

73. Wang X. H., Wang B. Z. and Chen K. J. Compact broadband dual-band bandpass filters using slotted ground structures. Progress in Electromagnetics Research 2008;82 (151-166.

74. Chen C. Y. and Hsu C. Y. A simple and effective method for microstrip dual-band filters design. IEEE Microwave and Wireless Components Letters 2006;16 (5) 246-248.

75. Koziel S. and Bandler J. W. Space mapping with multiple coarse models for optimization of microwave components. IEEE Microwave and Wireless Components Letters 2008;18 (1) 1-3.

CHAPTER 3

A Survey on M2M Systems for mHealth: A Wireless Communications Perspective

Elli Kartsakli [1,*], Aris S. Lalos [1], Angelos Antonopoulos [1,2], Stefano Tennina[3], Marco Di Renzo [4], Luis Alonso [1] and Christos Verikoukis [2]

[1] Department of Signal Theory and Communications (TSC), Technical University of Catalonia (UPC-BarcelonaTECH), C./ Esteve Terradas 7, Castelldefels 08860, Barcelona, Spain

[2] Telecommunications Technological Centre of Catalonia (CTTC), Av. Gauss 7, Castelldefels 08860, Barcelona, Spain

[3] WEST Aquila slr, Via Giovanni Gronchi 18, L'Aquila 67100, Italy

[4] Laboratory of Signals and Systems (L2S), CNRS - SUPELEC - Univ. Paris-Sud XI, 3 Rue Joliot-Curie, 91192 Gif sur Yvette, France

ABSTRACT

In the new era of connectivity, marked by the explosive number of wireless electronic devices and the need for smart and pervasive applications, Machine-to-Machine (M2M) communications are an emerging technology that enables the seamless device interconnection without the need of human interaction. The use of M2M technology can bring to life a wide range of mHealth applications, with considerable benefits for both patients and healthcare providers. Many technological challenges have to be met, however, to ensure the widespread adoption of mHealth solutions in the future. In this context, we aim to provide a comprehensive survey on M2M systems for mHealth applications from a wireless communication perspective. An end-to-end holistic approach is adopted, focusing on different communication aspects of the M2M architecture. Hence, we first provide a systematic review of Wireless Body Area Networks (WBANs), which constitute the enabling technology at the patient's side, and then discuss end-to-end solutions that involve the design and implementation of practical mHealth applications. We close the survey by identifying challenges and open research issues, thus paving the way for future research opportunities.

KEYWORDS

Machine-to-machine (M2M); mHealth; eHealth; healthcare; WBAN; end-to-end communications; IEEE 802.15.6

1. INTRODUCTION

Machine-to-Machine (M2M) communications is an emerging technology that envisions the interconnection of machines without the need of human intervention. The main concept lies in seamlessly connecting an autonomous and self-organizing network of M2M-capable devices to a remote client, through heterogeneous wired or wireless communication networks. An intelligent software application is usually employed at the remote client to process the collected data and provide the end user with a set of smart services and a practical interface. Although the idea of telematics and telemetry applications is not new, the widespread use of Internet, along with the trend for ubiquitous connectivity, especially via wireless communication systems, have placed M2M systems on the spotlight of attention for both academia and industry.

The increasing interest on M2M communications poses significant challenges that need to be met. A key issue to be handled is the large number of devices that must be supported in an M2M network, since market predictions estimate that the number of M2M-enabled devices with Internet connectivity will reach up to 50 billion by the end of 2020 [1]. Regardless of the exact figures, the growth rate is impressive, and major efforts are required to provide scalable solutions that support the increasing number of devices with diverse characteristics and requirements. Another challenge stems from the multitude of technical solutions that can be employed in M2M systems. Depending on the application deployment, different approaches may be adopted for the interconnection of M2M devices, such as wired or wireless technologies, short-range or long-range communications, and solutions based on existing open communication standards or proprietary technologies.

The above challenges stress the imperative need for standardization of M2M communications [2]. To this direction, the European Telecommunications

Standards Institution (ETSI) has established the M2M Technical Committee which aims to provide an end-to-end view of M2M standardization, focusing on the interoperability of M2M devices with existing standards. In July of 2012, ETSI and six other major standards development organizations (ARIB and TTC of Japan, ATIS and the TIA of the USA, CCSA of China, and TTA of Korea) joined their efforts in the one M2M initiative, under the goal of creating a single universal standard for M2M communications [3]. This global standardization effort is crucial to enable the integration of heterogeneous technologies in order to achieve seamless end-to-end connectivity, removing potential barriers to market growth.

The penetration of M2M solutions for monitoring and remote control in a wide range of markets, including industrial automation, security and surveillance, smart metering, energy management, and transportation, generates great business opportunities. The application of M2M enabling technologies to the healthcare sector, in particular, is expected to be one of the major M2M market drivers: market projections forecast that more than 774 million health-related devices with M2M connectivity will be available by 2020, yielding a total revenue of 69 billion euros in that year [4].

The use of mobile and wireless communication technologies to facilitate and improve healthcare and medical services, referred to by the term mHealth, is bringing a shift to healthcare delivery. The mHealth field can be considered as a subset of telemedicine, which is a broad term referring to the delivery of healthcare services between at least two remote locations, involving the communication between healthcare providers or the interaction between a healthcare provider and a patient [5]. Telemedicine can transform healthcare delivery across several medical disciplines, from cardiology and surgery to ophthalmology and psychiatry. The potential mHealth application scenarios are numerous, including the active management of diseases such as diabetes (e.g., by measuring blood sugar levels and controlling the insulin dosage accordingly), the support for independent aging to the elderly (e.g., by tracking their medication intake and their activity level) and the monitoring of personal fitness activities to improve health and well-being (e.g., by logging health and fitness indicators during workouts) [6]. As the sensing technology advances, future mHealth applications are expected to emerge [7,8], such as cancer detection (through the monitoring of cancer cells), management of heart condition (through the monitoring of biochemical markers) and vision enhancement (through implanted retina prosthesis chips), to name a few. Overall, mHealth can offer significant benefits for both patients and healthcare providers, ensuring enhanced quality, efficiency, flexibility and cost reductions in healthcare delivery.

The M2M paradigm in the context of mHealth involves the use of appropriate sensor devices on patients. The medical sensors, placed in the vicinity of, or inside, the human body, are usually interconnected through a short-range wireless technology, thus forming a Wireless Body Area Network (WBAN). An M2M-enabled gateway node collects all the sensory data from the WBAN and forwards them to a remote online server, where processing and integration with medical-related software applications take place. The connection of the gateway to the Internet is generally based on long-range communication access technologies for Wireless Local/Metropolitan Area Networks (WLANs/WMANs). The M2M technology paves the way for new possibilities for mHealth applications, by enabling the remote monitoring of vital signals [9], the early detection of critical conditions and the remote control of certain medical treatments [6]. The collection of large amount of data and their timely delivery to the healthcare providers in an unobtrusive way for the patient and without the need for human interaction can significantly facilitate the management of chronic diseases and speed up the diagnosis of critical conditions.

In the recent years, the research community has been motivated by the diversity of applications, the promising benefits and the potential market opportunities of mHealth M2M solutions. The main technological challenges for M2M communications, the most representative usage models and the status of global standardization efforts are discussed in [10]. Focusing on the emerging M2M technologies for mHealth applications, a technical discussion on the communication network design is given in [11], but generally most of the related work adopts a high-level approach. In [12], some interesting challenges from the network perspective are identified, while interoperability issues and recent standardization efforts are presented in [13]. On a different level, a lot of research activity has been focused on the body area domain, on the design of medical sensor devices [14] and on the main advances and challenges in the field of WBANs [15–18].

However, recent important work in the rapidly growing field of WBAN communication technologies have not been included in the aforementioned works, while the end-to-end holistic M2M approach is barely considered. In this context, this paper aims to provide a comprehensive survey on M2M systems for mHealth applications from a wireless communication perspective. This paper discusses many aspects of the M2M ecosystem in the healthcare domain, however, the main novel contribution is threefold:

- a systematic review of existing physical (PHY) layer technologies for WBANs, which, to our knowledge is not available in the literature,
- an up-to-date survey of the key Medium Access Control (MAC) layer design approaches to tackle the specific challenges of mHealth applications in WBANs, and
- an end-to-end perspective of M2M systems for mHealth, focusing on the integration and convergence of different communication technologies employed in WBANs and long-range communication networks and the presentation of testbed implementations.

The remainder of this paper is organized as follows. First, the high-level ETSI architecture for M2M systems is described in Section 2, whereas the main design challenges for end-to-end mHealth applications are discussed in Section 3. Section 4 is dedicated to the most recent developments in wireless technologies for WBANs. After a brief description of the IEEE 802.15.6 standard for WBANs, the focus is laid on channel modeling, PHY technologies and MAC strategies. Then, Section 5 is devoted to end-to-end connectivity, focusing on integration challenges, existing testbeds and current research projects. Finally, some concluding remarks are given in Section 6.

2. M2M NETWORK ARCHITECTURE

In the recent years, ETSI has been actively engaged in the development of a standard for M2M systems, with the objective to ensure interoperability between the diverse M2M components and the already existing technologies. To this direction, ETSI proposes a high-level horizontal architecture, dividing the system into three domains: (i) the device and gateway domain where the M2M devices communicate with a gateway through short-range area networks; (ii) the network domain that connects the gateway to the applications through long-range access and core communication networks; and (iii) the application domain where various application services are defined depending on the use case [19].

Figure 1 illustrates the architecture of a wireless M2M system for mHealth applications. The ETSI framework consists of five key elements as described below [19]:

- The M2M devices, which are devices capable of transmitting data autonomously or after receiving a data request. In the context of mHealth applications, the M2M devices are principally low-power medical sensors and actuators, placed close to or inside the patient's body (wearable or implanted sensors, respectively). Each device has an embedded wireless communication module, enabling their interconnection into a short-range network, as described in continuation.

- The M2M Area Network, also known as capillary network, which is the short-range network interconnecting the M2M devices and providing a link to the M2M gateway. Focusing on mHealth scenarios where wireless connectivity is considered among the body sensors, WBANs are the enabling technology for the M2M area network. Hence, in the rest of this paper, the terms M2M area network and WBAN will be employed interchangeably. IEEE 802.15.6 [20] is an emerging communication standard developed specifically for WBANs and suitable for healthcare applications. Other candidate technologies for the capillary network include ZigBee and IEEE 802.15.4, Bluetooth, and Bluetooth Low Energy. Proprietary solutions are also available, such as ANT, Sensium, Zarlink and Z-Wave.

- The M2M Gateway, which acts as a proxy between the M2M devices (interconnected through the WBAN) and the network domain. Practically, the gateway must be a portable device with advanced processing capabilities and multiple radio interfaces, able to operate in technologies employed by both the WBAN and the communication network. Typical example of M2M gateways include smartphones, Personal Digital Assistants (PDAs) and smart watches.

- The M2M Access Communication Network, which connects the M2M gateway to the Internet, eventually reaching the M2M application server. The communication network can be broken down to the access network and the core network. Common wireless technologies employed by the access network include Long Term Evolution (LTE), WiMAX and IEEE 802.11 WLAN. Core network technologies provide IP connectivity and include 3GPP, 3GPP2 and ETSI TISPAN core network systems.

- The M2M Application Server, which is the middleware layer that provides data to the specific business applications.

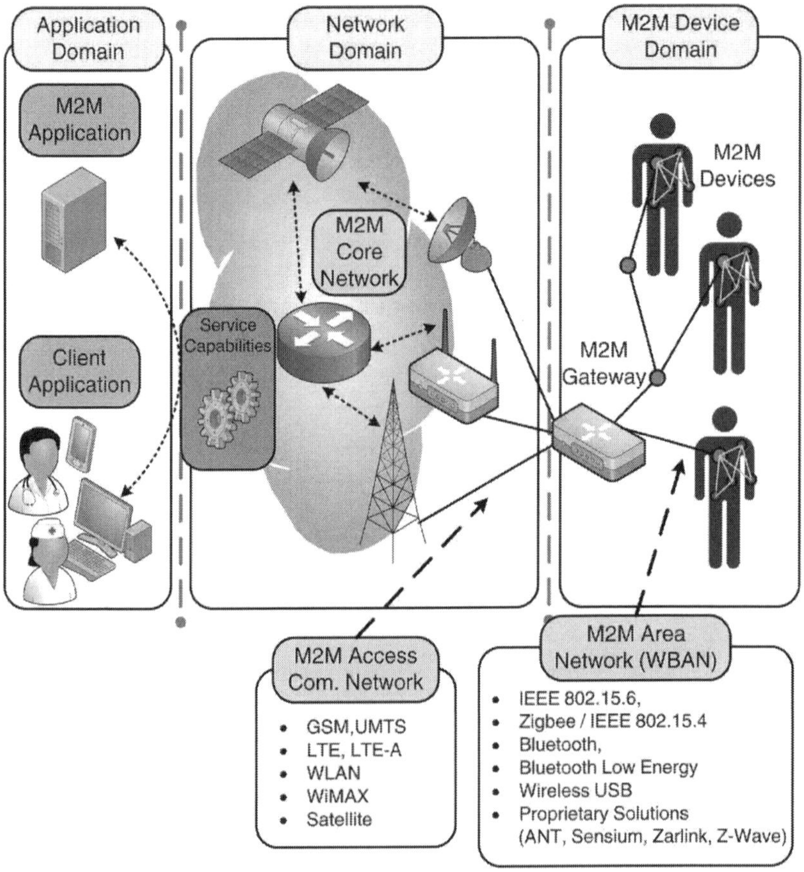

Figure 1. Simplified Machine-to-Machine (M2M) architecture for wireless connectivity in mHealth scenarios.

3. DESIGN CHALLENGES FOR MHEALTH APPLICATIONS

The previously described M2M system architecture can be employed in the context of wireless mHealth applications, aiming to form a communication bridge between the patients and the healthcare providers. At the patient's side, we consider WBANs, defined as low-power short-range networks deployed in the vicinity of the human body, as the prevalent technology for the M2M area (or capillary) network. The nature of WBANs poses inherent limitations not present in regular sensor networks, due to the specific characteristics and requirements of the employed M2M sensor devices. Furthermore, to provide end-to-end connectivity, the integration of different medium and long range technologies must be ensured, guaranteeing reliable and secure communication. The key design characteristics and challenges that must be taken into account at different stages of the M2M system, in order to design efficient techniques and communication protocols are summarized next [15,16,21,22]:

- Heterogeneous devices and traffic: In mHealth applications, the M2M devices are usually medical sensors that can be either attached on (wearable or on-body), or placed inside (implantable or in-body) the human body. These devices are capable of establishing wireless communication links, in order to enable continuous patient monitoring and provide real-time feedback to the responsible healthcare provider. Based on their operation, they can be classified as sensors or actuators. Sensors are used to measure external (e.g., motion, location, environmental temperature, etc.) or internal (e.g., heart beat, body temperature, muscle movement, brain activity, etc.) parameters of the human body. Actuators perform specific actions according to data received from the sensors or inserted by the end user (e.g., a pump for the administration of insulin based on blood glucose measurements). Unlike typical sensor networks, the number of devices in a WBAN is relatively small (i.e., usually restricted within the range of 20–50), with each device having a unique function. Table 1 contains a list of commonly employed sensors, emphasizing their diverse communication characteristics and requirements. As observed from the table, different types of data must be supported, ranging from plain messages to real-time audio and video content and continuous waveform signals, such as Electrocardiogram (ECG), Electromyograph (EMG), etc. Consequently, WBANs must be able to handle heterogeneous traffic and support a diverse range of transmission rates. Scalability is also a key feature, enabling the seamless addition or removal of nodes without affecting the network's operation.
- Wireless propagation characteristics: The propagation characteristics for most well established wireless technologies (e.g., cellular, WLANs, etc.) have been extensively studied. In the context of sensor networks, several studies have been made with respect to the signal propagation in different communication media [24]. Nevertheless, wireless communication close or inside the human body introduces new challenges due to the different

propagation characteristics of the body area environment. Even though the distances between the nodes are small, WBAN links suffer from high propagation losses, mainly due to user activity, in the case on on-body communications, and power absorption by the human tissue, in the case of implants. In addition, patient mobility and frequent changes in body posture can affect the quality of the wireless link.

- Energy efficiency: Even though energy efficiency is a desirable feature in all parts of an M2M communication system, it becomes of paramount importance in WBAN environments. To enhance comfort and unobtrusive wearability, body sensor devices must have a very small form factor, which unavoidably has a direct impact on the size and the capacity of the available energy sources (i.e., the batteries). Furthermore, there is an imperative need for prolonged battery lifetime, sometimes in the order of years, since battery replacement in the context of healthcare applications is not a trivial issue, especially in the case of implantable devices. Hence, efficient transceiver design, to ensure low-power consumption, and energy-aware communication protocols are essential.

- Quality of Service (QoS): The QoS requirements of an M2M mHealth system strongly depend on the application, given the very heterogeneous traffic characteristics that range from low-rate, predictable traffic to bandwidth-hungry real-time applications. Nevertheless, low end-to-end delay can be considered the most stringent QoS requirement in mHealth scenarios, especially when real-time monitoring is involved. As reported in [8], latency in medical applications should not exceed 125 ms, whereas bit error rates in the range of 10^{-3} to 10^{-10} should be supported, depending on the application. In addition, a critical issue in patient monitoring is the handling of alarm messages, generated by a sensor device when, for example, a sampled value exceeds some predefined limits. These unpredicted uplink data must be guaranteed timely delivery, since they can be closely associated to urgent, life threatening conditions for the patient. Typically, applications are classified into different service classes, based on their particular QoS requirements. Then, advanced scheduling and routing schemes are employed to provide different levels of priority to each service class.

- Reliability: Closely connected to QoS guarantees, reliability is another important issue in mHealth applications. We define reliability as the guaranteed end-to-end delivery of the transmitted data, from the patient to the medical personnel, or vice versa. The most vulnerable part of the M2M architecture in terms of reliability is the patient-side WBAN, due to the error-prone WBAN channels, which must take into account the particular propagation characteristics of implanted and on-body wireless links and patient mobility.

- Interference and coexistence: WBANs operate on very low transmission power, given the short-range connectivity requirements and the strict regulations on the acceptable Specific Absorption Rate (SAR) limits (to reduce the risks from radiation exposure). Hence, on one hand, interference caused by WBANs is relatively low. On the other hand, interference tolerance and coexistence with other wireless systems can be a challenging issue, especially for WBANs operating at the industrial, scientific, and medical (ISM) frequency band and deployed in environments with many interfering sources.

- Topology: Typically, the most common WBAN configuration is the star topology, where all the sensor devices are connected to the WBAN coordinator via a direct link. However, it is also possible to use part of the deployed sensors as relays [25], to improve connectivity in case of poor quality direct links, thus enhancing reliability [26,27] and energy efficiency [28]. In the M2M access communication network, a point-to-point medium/long-range connection between the WBAN gateway and the core network is usually assumed (e.g., through a WLAN access point or an LTE base station), without excluding the possibility of more complex topologies. For instance, it is possible to exploit the presence of multiple WBANs [29] or ambient sensor network deployments [30] in order to establish multihop links, enabling the design of efficient routing and cooperative schemes.

- Security: A key challenge of mHealth applications is to guarantee of the confidentiality of medical information, which faces additional threats due to the M2M architecture, mainly due to: (i) the wireless transmission of medical-sensitive data; and (ii) the data storage in multiple locations, both locally and in remote servers and devices. Hence, a need for a holistic approach is generated in order to ensure secure M2M communication [31]. To this direction, new schemes have to be designed, taking into account the specific characteristics of different technologies, in order to create interoperable and technology-independent protocols and security mechanisms.

- Technology integration: The wireless technologies employed in the different stages of the M2M system architecture have diverse characteristics and challenges that are often studied separately. Nevertheless, the integration of these technologies into a unified mHealth application is an open issue that must be considered carefully. Therefore, in order to guarantee end-to-end QoS, scalability and ubiquitous connectivity, it is important to adapt the access communication technologies, such as LTE, WiMAX and IEEE 802.11 WLAN, to the characteristics of WBANs, taking into account the requirements of pervasive mHealth applications.

Table 1. Technical characteristics of selected Wireless Body Area Networks (WBAN) sensors for mHealth applications [18,23].

Sensor	Battery Lifetime	Data Rate	Criticality	Application
Deep brain stimulation	>3 years	1 Mbps	High	Therapeutic benefits for Parkinson's disease, chronic pain, tremor, and dystonia
Hearing aid	>40 h	200 kbps	High	Sound amplification
Accelerometers/Gyroscopes	>1 week	1 Mbps	Low	Measurements on motion detection, acceleration and angular velocity
Pulse oximeter (SpO_2)	>1 week	2 kbps	Low	Measurement of hemoglobin oxygen saturation
Capsule endoscope	>24 h	1 Mbps	High	Imaging of the digestive tract
Temperature	>1 week	2.4 bps	Low	Body or environmental temperature measurements
Electrocardiogram (ECG)	>1 week	9.6 kbps	Medium	Heart waveform characteristics
Electromyograph (EMG)	>1 week	100 kbps	Medium	Muscle movement
Gate/Falls	>1 week	250 kbps	High	Fall detection
Electroencephalo-gram (EEG)	>1 week	100 kbps	High	Brainwave activity
Video/Medical imaging	> 12 h	<10 Mbps	High	Digital video transmission

4. WBANS: THE KEY M2M AREA NETWORK TECHNOLOGY FOR MHEALTH APPLICATIONS

As seen in the previous section, many of the challenges present in M2M mHealth systems are related to the area network domain, thus stressing the important role of WBAN technologies. Hence, in this section, we aim to provide an overview of the research work on some key aspects of WBAN design. First, we provide a concise description of IEEE 802.15.6, which is the latest IEEE standardization activity targeted specifically to WBANS. Then, we describe the main channel modeling methods and propagation characteristics for body area communications. In continuation, we focus on different PHY layer technologies that can be employed in WBANs. Finally, this section ends with an up-to-date survey of the most representative MAC mechanisms designed in the context of mHealth applications in WBANs.

4.1. IEEE 802.15.6 Standard for WBANs

In an effort to tackle the specific requirements of WBANs, the IEEE 802.15.6 standard has recently been issued for short-range wireless communications in the vicinity of, or inside, the human body. A key characteristic of IEEE 802.15.6 is that it supports operation at very low transmission powers, as well as macroscopic and microscopic power management through hibernation and sleep modes, respectively, in an effort to increase battery lifetime and comply with the safety regulations limits on the maximum acceptable absorption level (i.e., SAR) for in-body communications. It also provides QoS guarantees for the prompt delivery of alarms in emergency situations and employs robust security mechanisms to provide privacy and confidentiality protection of the medical data.

The standard provides specifications for the PHY and the MAC layers. With respect to the PHY layer, three different technologies are supported: (i) the Narrowband (NB) PHY, which introduces low control overhead, very low peak power consumption and robustness against interference; (ii) the Ultra-Wideband (UWB) PHY, based on a technology for transmitting information over a large bandwidth, offering high performance, robustness, low complexity and ultra-low power operation; and (iii) the Human Body Communications (HBC) PHY, which uses the human body as a means of propagation for the data transmission, through the galvanic coupling of signal currents.

Operation at multiple frequency band is supported, starting from the Medical Implant Communications Service (MICS) band of 402–405 MHz, reserved for medical implant communication, up to the unlicensed 2.4 GHz ISM band. Table 2 summarizes the frequency bands supported by each PHY layer technology, as well as the range of achievable transmission rates.

Table 2. IEEE 802.15.6 frequency bands and supported rates.

PHY Layer	Frequency Band (MHz)	Bandwidth (MHz)	Available Channels	Rates (kbps)
Narrowband (NB)	402–405 (MICS)	0.3	10	75.9–455.4
	420–450	0.32	12	75.9–187.5
	863–870	0.4	14	101.2–607.1
	902–928 (ISM)	0.4	60	101.2–607.1
	950–958	0.4	16	101.2–607.1
	2,360–2,400	1	39	121.4–971.4
	2,400–2,483.5 (ISM)	1	79	121.4–971.4
Ultra-Wideband (UWB)	3,000–5,000	499.2	3	Non-coherent: 394.8–12,636 Differentially coherent: 487–15,600
	6,000–10,000	499.2	8	FM: 202.2
Human Body Communications (HBC)	21 (center frequency)	5.25	1	164.1–1,312.5

MICS: Medical Implant Communications Service; ISM: Industrial, Scientific and Medical; FM: Frequency Modulation.

With regard to the MAC layer, IEEE 802.15.6 defines eight levels of user priorities, with level 0 corresponding to the lowest priority class and level 7 being assigned to the highest priority traffic for emergency situations or medical implant event reports. The WBAN operates in an extended star topology, with all nodes connected directly, or through a single relay node, to the coordinating node, denoted as the hub. The standard considers both contention-based and contention-free channel access.

Two random access schemes for contention-based access are supported: (i) Carrier Sense Multiple Access with Collision Avoidance (CSMA/CA); and (ii) slotted Aloha. In CSMA/CA, a node has to sense the medium idle before transmitting and executes a backoff mechanism to reduce the collision probability and resolve contentions in case of simultaneous transmission attempts. In slotted Aloha, time is divided into slots and nodes attempt transmission at the beginning of each slot with a given probability that depends on the user priority.

The standard also supports three contention-free access modes that require centralized control from the WBAN hub: (i) improvised access, which is a polling scheme for immediate of future allocation for both uplink and downlink; (ii) scheduled access, which are periodic allocations for uplink, downlink or bidirectional transmissions that are negotiated between the nodes and the hub during the association phase; and (iii) unscheduled access, which is a best-effort version of the scheduled access.

As far as security is concerned, the standard defines three different connectivity levels: (i) unsecured communication; (ii) authentication only; and (iii) authentication and encryption.

4.2. WBAN Channel Modeling

Accurate and reliable channel models are crucial for the design and evaluation of transmission and reception techniques employed at the PHY layer. Hence, considerable efforts have been made to define the characteristics of the body area propagation environment through extensive measurements and simulations. In this section, we present the most important results related to the channel modeling for on-body, in-body and HBC communications in WBANs, along with the different factors that affect the propagation characteristics.

4.2.1. On-Body Radio Propagation Model

This section summarizes the most important findings related to the path loss and the small-scale fading of on-body radio channels.

Path loss Model:*According to [32–35], propagated waves are more likely to diffract around the human body, rather than penetrate it. Hence, traditional formulas for wireless communications can be applied, such as the Friis equation that caclulates the path loss as a function of the distance d [36]:*

$$\text{Path loss} \sim G_r G_t \left(\frac{\lambda}{4\pi}\right)^2 \left(\frac{1}{d}\right)^n \tag{1}$$

where G_t and G_r are the respective transmitter and receiver antenna gains, λ is the wavelength, and n is the path loss exponent that depends on the propagation environment.

In free space propagation environments, the path loss exponent is equal to n = 2. In the context of WBANs, however, n has a higher value, depending on the scenario. In particular, when the transmit and receive antennas are placed at the same side of the body (Line-of-Sight-LOS situation), it has been estimated that the values of n range between 3 and 4, depending on the position of the device (i.e., the path loss on the arm is lower that the one on the trunk). Then, in Non-LOS (NLOS) cases that occur when the antennas are placed at different sides of the body, n assumes higher values, between 5 and 7.4 [33,35].

It is important to mention that in the context of WBANs, frequency does not significantly affect the path loss [37]. On the other hand, the current user activity and the antenna height (defined as the distance from the body) are the two major contributors to the path loss characteristic of wearable WBANs. Regarding the former, the impact of user activity (i.e., walking) on the path loss has been experimentally studied in [38]. The obtained results show that the subject's mobility causes significant deviations in the path loss, with respect to the average path loss values obtained when the subject remains static. Concerning the antenna height, the authors in [39] claim that the closer the antenna is to the body, the higher is the path loss. A difference of more than 20 dB can be obtained by comparing the path loss for antennas placed at 5 mm and 5 cm from the body surface.

Small-scale fading:*Small-scale fading is different in on-body WBANs with respect to other wireless environments. Several distributions have been proposed to model small-scale fading in different frequency bands [15,40]. However, traditional Rayleigh and Rician distributions have been proven inefficient, while the lognormal distribution seems to be the best fit [41]. In addition, measurements in [40] have indicated that the multipath components can be divided into two clusters. The first cluster is due to the initial wave diffraction around the body and the second is due to reflections on the ground. The correlation between the bins of the two clusters is very weak, and therefore the bins from the different clusters can be considered as statistically independent. On the other hand, the different delay bins of the same cluster are* *highly correlated. This can be explained by considering that when the path length is very short, the multipath components have overlapping path trajectories. In a real situation, however, the reflections from the scatterers in the surrounding environment [35] should be also taken into account.*

In conclusion, the recommended body area channel model for wearable WBANs should consider a fixed two cluster mutlipath model containing: (i) a dominant direct component; (ii) a set of early reflected waves adding (possibly distructively) to the dominant wave; and (iii) a set of excessively delayed waves, adding incoherently to the dominant wave. The dominant wave amplitude is determined by an exponential path loss and a log normal fading coefficient that is strongly correlated with the fading coefficients of the early delayed waves.

4.2.2. In-Body Radio Propagation Model

As mentioned in the previous section, communication of medical implantable devices is restricted to the MICS frequency band of 402–405 MHz. In this frequency range, the human body can be considered as an inhomogeneous medium consisting of multiple tissue layers with different electrical characteristics. Therefore, it behaves as a communication channel where losses are mainly due to the power absorption by the different tissue layers.

Radio propagation through human tissue has been extensively studied in the literature ([42–46]) and depends on many parameters, including, among others, sex, age, location of the sensor in the human body and general posture of the body. The derivation of an exact expression for the in-body propagation model is extremely difficult. Instead, the SAR level is employed as a standard measure of how much power is absorbed by the tissue. Compared to the free space propagation, an additional 30–35 dB path loss is noticed at small distances. Measurements in [44–46] demonstrated that direct communication between implanted nodes suffers from a lower path loss with respect to the link establishment between an implant and a node in body surface. Finally, it has been shown that the shape and the position of the patient's body has a significant impact on the radiation pattern of an implanted radio transmitter [47].

4.2.3. HBC Propagation Model

Apart from the propagation of radio waves in human body, WBANs support HBC, based on data transmission by capacitive and galvanic coupling, as shown in [48–51]. The HBC propagation characteristics are mainly determined through measurements aiming to study the electrical behavior of the human body. In [48], it has been shown that the transmitted signal level experiences high variations at different locations of the body. For example, short-range links between devices placed on the thorax or on the limbs have shown excellent transmission characteristics, whereas high attenuations that prohibit long-distance transmissions have been measured for extremities and joints.

4.3. PHY Technologies for WBANs

PHY layer technologies determine the way in which the source data is converted into electrical signal for transmission. In order to increase the energy and bandwidth efficiency, as well as the transmission reliability, a series of signal processing functions must be carried out, including channel coding and decoding, modulation and demodulation, filtering and frequency conversion, synchronization, channel estimation and equalization. All these operations are executed at both the transmitter and the receiver of a digital communication system, which are typically implemented together in a single device, referred to as the transceiver. This section aims to provide a survey of the main challenges in the optimization of different transceiver parts in WBANs for mHealth applications, filling the respective gap that, to our knowledge, exists in the literature.

In accordance to the PHY layer technologies supported by the IEEE 802.15.6 specification, as mentioned in Section 4.1, we divide this section into three parts, devoted to NB, UWB and HBC transceivers, respectively. Particular emphasis will be placed on UWB transceivers, since it is a promising technology for WBAN applications, due to its high energy efficiency and low complexity.

4.3.1. NB Technologies

NB transceiver technologies have been traditionally employed in short and medium range technologies (e.g., Zigbee, Bluetooth, IEEE 802.11 WLANs) and are widely used in commercially available products. In WBANs, NB systems can operate at different frequency bands (Table 2) and are suitable for both on-body and in-body communications. Most existing works on NB transceiver design are focused on the implementation of energy-efficient modulation and demodulation schemes, and their contributions are summarized in the rest of this section.

Two main classes of NB receivers for on-body WBANs may be found in the literature: (i) ZigBee receivers, equipped with quadrature mixers that support Quadrature Phase Shift Keying (QPSK) modulation. The power expensive Phase-Locked Loops (PLLs) and the quadrature oscillators decrease the energy efficiency of NB receivers with respect to UWB or HBC receivers; (ii) Receivers that use regenerative positive feedback to achieve gain, and non-coherent detection to improve energy efficiency. The modulation schemes applied in this case are simple amplitude-based signaling schemes such as On-Off-Keying (OOK). The main drawback of these receivers is the poor sensitivity to RF signals, that can cause oscillation. When oscillating, not only is it impossible to receive any signal, but those oscillations are driven against the antenna and retransmitted into the air, causing interference to nearby receivers.

In the case of in-body communications, given that low-power consumption is one of the key requirements, energy-efficient modulations with small envelope variations are favorable. However, when high data rates must be employed, it is of great importance to choose a high-sensitivity modulation and demodulation technique in order to overcome the strong attenuation caused by the human body. The modulation schemes that have prevailed in NB systems for mHealth applications are $\pi/4$-shift QPSK, offset QPSK, FSK and MSK (Minimum Shift Keying). Oh et al. [52] propose a new type of PSK modulation for high data rates, called Phase Silence Shift Keying (PSSK), which is more bandwidth-efficient than orthogonal modulation schemes such as OOK. Under an Additive White Gaussian Noise (AWGN) channel model with path loss for the human body, the authors in [52] show that the BER of an 8-PSSK is lower than that of various sinusoidal carrier-based modulation schemes. The authors in [53] propose a simple ultra-low power hardware design for an FSK/MSK direct modulation transmitter in medical implant communications. Finally, an extended protocol that uses a rateless code with FSK modulation in order to overcome the reliability and power cost concerns in implanted sensors has been proposed in [54].

4.3.2. UWB Technologies

UWB transceiver technologies, mainly used for on-body communications, are characterized by very low radiated power (i.e., −41.3 dBm/MHz), low implementation complexity, and harmonic coexistence with other existing wireless technologies. Hence, there is an increasing interest in the use of UWB transceivers in biomedical scenarios.

Based on the method employed to form the carrier signal, the UWB technologies are classified into three groups: (i) Impulse Radio UWB (IR-UWB); (ii) Frequency Modulated UWB (FM-UWB); and (iii) Multi-Band Orthogonal Frequency Division Multiplexing UWB (MB-OFDM UWB). A brief description and related work on each technology are presented in continuation.

IR-UWB : *IR-UWB employs an ultra-short time pulse (in the order of ns) and, therefore, occupies a wide band in the frequency domain. IR-UWB signals are characterized by: (i) very wide bandwidth that can exceed 1 GHz; (ii) very low power spectral density; and (iii) a relatively simple and inexpensive architecture for both transmitter and receiver. A simple transceiver structure for IR-UWB signals is shown in Figure 2.*

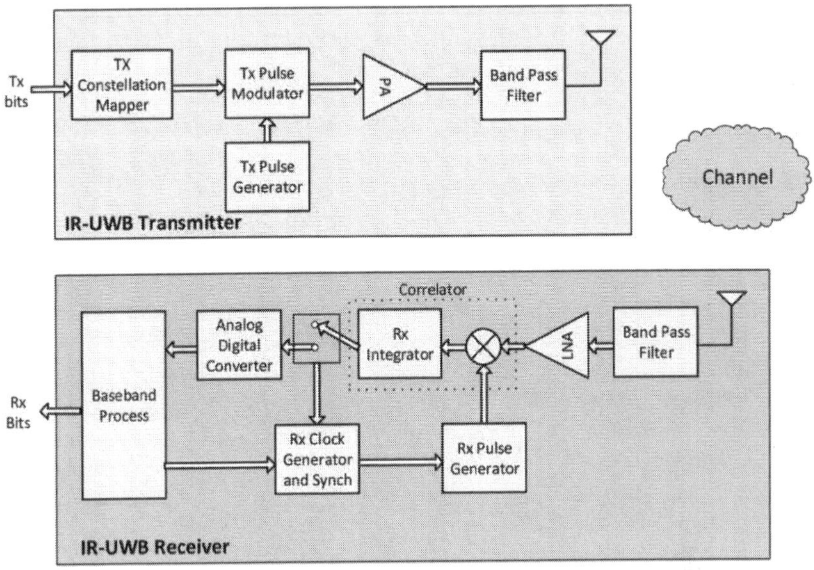

Figure 2. Impulse Radio Ultra-Wideband (IR-UWB) Transceiver.

IR-UWB can support many of the well known modulation schemes that are usually employed in NB systems [55], resulting to a wide range of achieved data rates. In order to demodulate IR-UWB signals, coherent, differentially coherent, and non-coherent detection schemes can be applied. A performance study of an IR-UWB system for WBANs is presented in [56], where the impact of two different modulation techniques on the Bit Error Rate (BER) of different on-body radio links is analyzed. The authors in [57] have studied different (coherent and non-coherent) UWB receivers, implemented according to the IEEE 802.15.4a requirements in different hospital environments, showing that both receivers yield equivalent performance in short-distance LOS links.

The authors in [58] have investigated the problem of pulse distortion due to the interaction between UWB electromagnetic pulses and the human body. By studying the performance of various detection algorithms in the presence of pulse distortion and receiver noise, it has been shown that non-coherent receivers, such as energy detectors and Transmitted-Reference (TR) receivers (In TR systems, two signals—a reference and a data signal—are transmitted per bit. The information is contained in the sign difference of the transmitted signals.), are suitable for WBANs, due to the fact that they do not require any explicit channel estimation (thus consuming less power and reducing complexity). In [59], two non-coherent modulation/detection pairs for IR-UWB systems with and without chirp pulse compression have been investigated. The authors strongly recommend the usage of chirp pulse compression in medical IR-UWB WBAN devices that are expected to use low data rates. Non-coherent detectors for Pulse Position Modulation (PPM) and TR Pulse Amplitude

Modulation (TR-PAM) are proposed in [60]. The authors verify via simulations that the knowledge of the average power delay profile at the receiver can substantially improve the system performance.

FM-UWB: *FM-UWB is a low-complexity, constant envelope, low data rate UWB communication technology with good interference robustness at very low data rates and lower power consumption with respect to IR-UWB. FM-UWB is typically an analog implementation of a spread-spectrum system using double FM: digital Frequency Shift Keying (FSK), followed by high modulation index analog FM to create a constant envelope UWB signal. A typical FM-UWB transceiver architecture is presented in Figure 3. The data signal modulates a low-frequency subcarrier using NB FSK techniques. Then, the subcarrier signal modulates the RF oscillator using analog FM with high modulation index. The receiver demodulates the FM-UWB signal without requiring local oscillator and carrier synchronization (as in the case of IR-UWB), which makes the device both simpler and cost-efficient.*

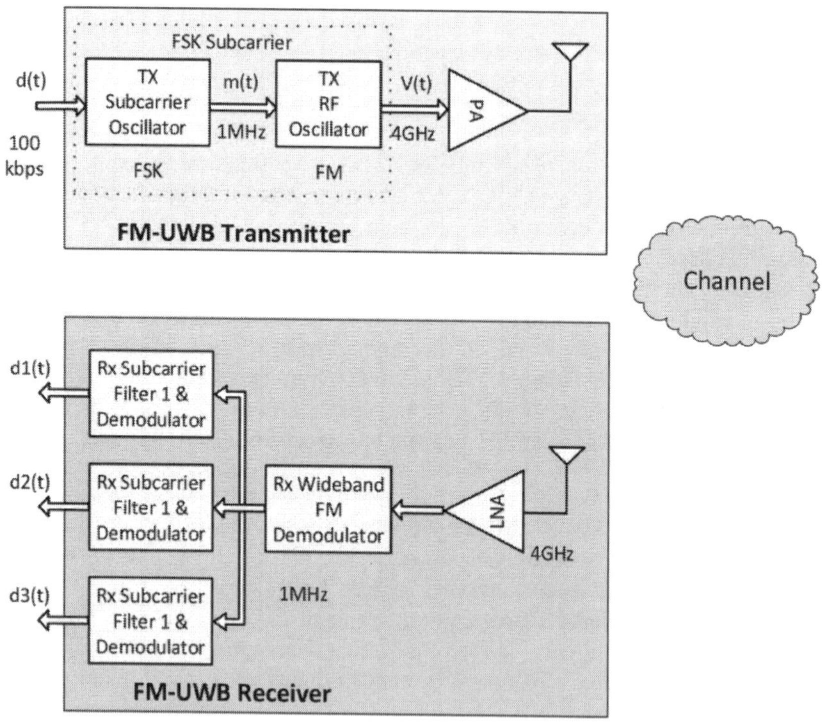

Figure 3. FM-UWB Transceiver.

The main characteristics of FM-UWB are: (i) simple and straightforward transmitter implementation; (ii) no need for carrier synchronization at the receiver; and (iii) system operation with a variety of antennas, adding robustness against NB jamming.

A novel system for the monitoring of vital signs based on the FM-UWB technology is presented in [61]. The authors study the error introduced by each block of the transceiver, in order to understand the achievable resolution under non-ideal working conditions. A comparison with an IR-UWB system is also conducted, showing that the FM-UWB system can achieve the same performance, working with lower power levels. The authors in [62] propose the implementation of a device which integrates both sensing and communication capabilities by using the same FM-UWB technology. Finally, the authors in [63] compare the performance of IR-UWB and FM-UWB transceivers in various channel models, showing that while the selection of the channel model has a has a major impact on the FM-UWB performance, the performance of the IR-UWB transceiver is not significantly affected.

MB-OFDM UWB: *MB-OFDM UWB is a technique based on the simultaneous transmission of data over multiple carriers spaced apart at precise frequencies. A typical architecture of an MB-OFDM UWB transceiver is illustrated in Figure 4, where the carrier and sampling synchronization is performed after the analog front end. In the specific architecture shown in the figure, time and frequency synchronization (carrier and sampling offset estimation and compensation) are performed in the time domain, while channel estimation and equalization are performed in the frequency domain. Nevertheless, different transceiver designs can be implemented, depending on the adopted algorithms for synchronization, channel estimation and signal detection.*

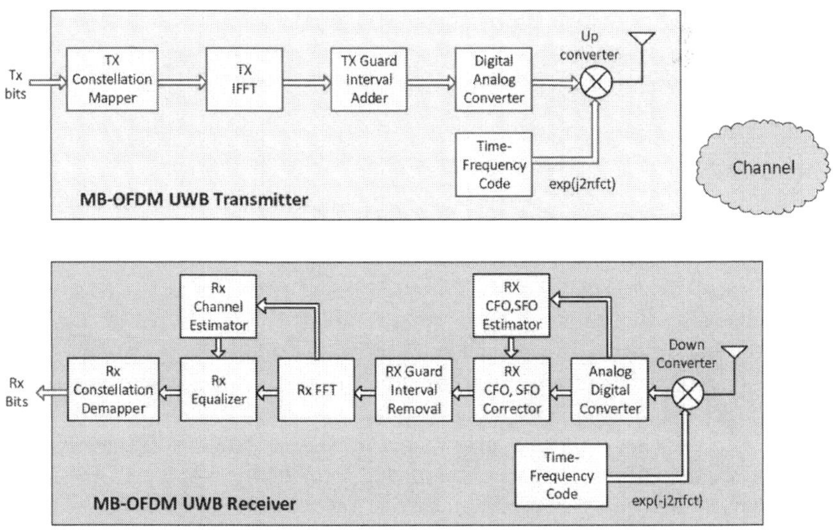

Figure 4. Multi-Band Orthogonal Frequency Division Multiplexing UWB (MB-OFDM UWB) Transceiver.

Compared to IR-UWB modulations, MB-OFDM UWB can make a more efficient use of the spectral resources. Moreover, a narrower sub-band bandwidth relaxes the requirement on the sampling rates of analog-to-digital converters, and consequently, facilitates digital processing. The use of OFDM makes the receiver spectrum-efficient and capable of better capturing multipath energy, compared to an equivalent single-carrier Rake receiver using the same total bandwidth. On the other hand, the drawbacks of an MB-OFDM UWB transceiver are that: (i) the transmitter is more complex because it requires an IFFT; (ii) the peak-to-average ratio may be slightly higher than that of the pulse-based multi-band approaches; and (iii) the generation of multi-tones requires high power consumption.

The authors in [64] provide a system-level analysis of MB-OFDM UWB radio systems, by calculating the BER of several on-body radio links measured in an indoor environment. Results demonstrate that the system performance is significantly affected by the position of the on-body sensors, as well as by the changes in body posture, highlighting the importance of considering these parameters when designing optimal MB-OFDM UWB radio systems for WBANs. A thorough investigation of key design issues for MB-OFDM UWB systems, such as FFT size, guard interval size and coding scheme, is provided in [65]. Finally, the NB interference impact on MB-OFDM-UWB systems has been examined both theoretically and by means of simulations in [66], revealing the importance of developing novel and low-complexity NB interference mitigation schemes for OFDM-UWB receivers.

Another big challenge in MB-OFDM UWB is that the constraint of ultra-low transmission power complicates channel estimation. Moreover, the heterogeneous environment of WBANs leads to a dense multipath wireless channel. The authors in [67] present an overview of channel estimation techniques for MB-OFDM UWB communications, highlighting their strengths and weaknesses, and providing useful suggestions for the adoption of these methods for WBAN channels.

4.3.3. HBC Technologies

Unlinke the other two WBAN PHY layer wireless technologies, HBC uses the human body as the communication medium. HBC transceivers enable low-power communication at low frequencies. Previous studies have shown that the optimal operation frequency range for body coupled communications ranges between 10–100 MHz [68,69]. The lower bound is imposed by propagation losses while the upper bound is determined by the fact that the RF energy is no longer confined to the body but is radiated into the environment. Therefore, HBC signals must be modulated to a carrier lower than 100 MHz, before being capacitively coupled to the body via electrodes on or near the surface of the skin. Recently published works on HBC transceivers [70,71] have shown energy efficiency in the order of 1 nJ/bit, data rates of 1 Mbps and receiver sensitivity of −60 to −70 dBm.

The low amount of power radiated outside the body by an HBC transceiver provides a layer of inherent security to the user, allowing spectrum reuse between adjacent users without requiring an explicit cellular structure. A major concern, however, is the interference cancellation due to the presence of high-powered television and radio stations that coexist in the same bandwidth. The authors in [70] adopt a cognitive FSK modulation to avoid frequencies with high interference, whereas the authors in [71] employ a high-input resistance at the receiver to reject offending spectrum above 30 MHz, and a correlation-based scheme to reject signals with low correlation with respect to the expected received signals.

An overview of the presented related work on PHY layer transmission techniques has been provided in Table 3.

Table 3. Overview of physical (PHY) layer transmission technologies [18,23].

WBAN PHY Layer Technologies		Overview of Proposed Schemes
NB Technology		[52] Novel modulation and demodulation scheme
		[53] Novel hardware design
		[54] Novel modulation and coding scheme
UWB Technology	IR-UWB	[56] Channel modeling and performance study
		[57] Comparative study of IR-UWB receivers (coherent and non-coherent)
		[58] Channel measurements and channel impact on performance and localization
		[60] Evaluation of novel receiver designs with channel state information knowledge
		[72] Design and evaluation of an UWB transmitter for low-power WBANs
	FM-UWB	[61] System design and comparison with IR-UWB
		[62] Integration of sensing and communication devices using FM-UWB to a single device
		[63] Performance study of DS-UWB and FM-UWB under two different channel models
	MB-OFDM UWB	[64] Channel modeling and system level evaluation
		[65] Receiver design and investigation of the impact of several design parameters
		[66] Impact of NB interference and a novel mitigation scheme
		[67] Overview of constant envelop and carrier-frequency offset mitigation techniques
HBC Technology		[70] Novel modulation and demodulation scheme
		[71] Transceiver design and simulation study
		[73] Baseband system design and simulation study

4.4. MAC Layer Design for WBANs

The main functions of the MAC layer include: (i) the access regulation to the shared wireless medium among the system nodes; and (ii) the scheduling of transmissions to achieve desired performance goals, such as throughput guarantees, time-delay constraints and QoS provisioning. In addition to these functionalities that are common to all systems, a MAC protocol suitable for WBANs must handle specific challenges, associated with the WBAN topology and the node constraints. Furthermore, the MAC protocol design has a significant impact on energy efficiency, QoS provisioning, reliability and interference, which, as mentioned in Section 3, are some key requirements of mHealth applications.

Several extensive surveys and comprehensive guidelines on MAC layer design for the wider category of sensor networks are available in the literature, such as the works in [74,75] and more recently in [76]. Even though these works can provide helpful insights for many of the challenges also faced in WBANs, especially in the design of low-energy MAC protocols, they are not specifically addressed to healthcare applications. Two overviews on WBAN-oriented MAC protocols have been published in 2010 [77,78], but are missing later advancements in this rapidly growing research field. Finally, more recent surveys on WBANs that include discussions on the MAC layer challenges can be found in [15,16,21,22].

In this section, we update and complement the existing surveys by identifying the main challenges and design approaches to tackle the aforementioned requirements of mHealth applications in WBANs. After a thorough study of the proposed MAC protocols in the literature, we provide a survey of the representative mechanisms employed for (i) energy-efficient channel access and synchronization; (ii) QoS provisioning; (iii) reliability; and (iv) interference mitigation and coexistence.

4.4.1. Energy-Efficient Channel Access and Synchronization Mechanisms

In an ideal energy-efficient MAC protocol, nodes would spend most of their time with their radio interface turned off (or in a low-consumption sleep mode) and wake up only to transmit or receive data in a collision-free way and without unnecessary listening times. However, in practice, significant energy is consumed in data collisions and idle listening periods, occurring when the nodes keep listening to the channel, usually after waking up from sleep mode [77]. To reduce power consumption, most WBAN-oriented protocols employ sleep modes. The challenge, however, is to design mechanisms that handle efficiently the periods during which the nodes remain awake. This is mainly achieved through channel access schemes that reduce data collisions and minimize listening times. Synchronization is also an important issue, since most energy-efficient access mechanisms require common time reference within the WBAN. The rest of this section summarizes the main mechanisms adopted in the literature with respect to these issues.

Collision-free channel access: *Many proposals guarantee collision-free data transmission by introducing low-cycle Time Division Multiple Access (TDMA)* *scheduling schemes where channel access is managed by a central coordinating node. An example of a straightforward TDMA implementation has been proposed in [79]. The proposed superframe structure starts with the transmission of a beacon by the coordinator, followed by time slots that are allocated to nodes according to a known and predefined schedule that matches their periodic traffic patterns. A more advanced TDMA-based solution for periodic traffic that takes into consideration packet retransmission due to channel errors has been proposed in [80]. In the proposed scheme, nodes are again granted a fixed slot allocation in every superframe, long enough to transmit a data packet and receive an acknowledgment (ACK) by the coordinator. In addition, some extra slots are reserved for packet retransmissions in case of channel errors. In both schemes, nodes wake up at predefined intervals to receive beacons and transmit data.*

Another mechanism for collision-free data transmission has been proposed in CICADA (Cascading Information retrieval by Controlling Access with Distributed slot Assignment) [81]. In CICADA, the network is organized in a tree topology, with the coordinator placed at the top. The protocol operation defines two cycles for control and data transmission, further divided into time slots. During the control cycle, a slot assignment scheme is forwarded from parent to child node, starting from the coordinator. During the data cycle, the nodes wake up during their allocated slots to transmit data towards the coordinator, while data is flowing upwards from the nodes at the bottom of the tree hierarchy towards the coordinator.

Hybrid channel access:*Pure TDMA-based strategies completely eliminate data collisions at the cost of reduced flexibility in scheduling. A more versatile approach is to adopt hybrid schemes that combine different mechanisms for channel access and data transmission. The main concept is to divide time into two phases: (i) a contention-based phase, during which the nodes compete for channel access, in order to inform the coordinator of their bandwidth requirements; and (ii) a collision-free phase, during which nodes can proceed to data transmission, after the resource allocation negotiations that took place in the contention-based phase.*

An example of a hybrid access mechanism has been implemented in BodyMAC [82]. The protocol introduces a superframe structure divided into three parts: a synchronization beacon, a downlink part reserved for the transmission of data from the coordinator and an uplink part for data transmitted by the nodes to the coordinator. The uplink is split into a Contention Access Part (CAP), during which the nodes can access the channel through CSMA/CA rules, and a Contention-Free Part (CFP), during which time slots are allocated dynamically to the nodes by the coordinator. In order to gain slot allocation, the nodes must notify their bandwidth requirements to the coordinator during the CAP period. Then, different bandwidth allocation schemes can be implemented for periodic or temporal slot assignments. Sleep mode is supported through a three-step handshake, initiated by the transmission of a sleep request packet by the node during the CAP. The coordinator acknowledges reception with an ACK packet and transmits a sleep mode schedule in the downlink period of the following superframe, containing the necessary sleep parameters.

A different implementation of the same principle is adopted in Distributed Queuing Body Area Network (DQBAN) [83]. The superframe structure in DQBAN is again divided into three parts: (i) the access part, further divided into a number of access minislots; (ii) the scheduling part, further dividing into a number of scheduling minislots; and (iii) the data part, in which collision-free data transmission takes place. Nodes compete for channel access by transmitting small control packets in randomly selected access minislots, and a collision resolution algorithm is employed to resolve any collisions among access requests. After a successful access request, nodes enter a transmission queue and are enabled to transmit their packet in the data part in a collision-free manner.

Finally, an energy-aware hybrid access scheme that exploits energy harvesting at the wireless nodes has been presented in [84,85]. The proposed scheme, HEH-BMAC combines two different access methods: (i) a reserved, collision-free polling access phase for nodes with high energy level and more urgent data; and (ii) a probabilistic random access phase for nodes with low energy level. The key novelty of this scheme is that the duration of the two phases is dynamically determined, taking into account the energy level of the body nodes.

Preamble sampling: *Another class of protocols adopts a technique often referred to as preamble sampling [76]. The main concept is that the nodes follow independent sleep patterns, spending most of their time asleep, waking up only for short periods of time. While they are awake, they listen to the channel for any beacons transmitted by the coordinator. If a beacon is received, they proceed to transmit their data, otherwise they return to sleep mode. Energy consumption can be reduced by minimizing the idle listening time (i.e., the time spent for waiting to receive a beacon). This can be accomplished by adapting the beacon transmission intervals to the sleep pattern of the nodes.*

A mechanism to deal with this issue is based on the negotiation of sleep pattern between the coordinator and the nodes. In [86], for example, a master-slave architecture is assumed and the master keeps track of the sleeping pattern of each slave node. The slave remains in a sleep mode until the predefined sleep time elapses, and then wakes up and waits to be polled by the master. A new sleep time is then defined and the process is repeated. In Traffic-adaptive MAC (TaMAC) [87], the coordinator determines the wake-up patterns of the nodes and stores them in a table. By consulting this table, the coordinator can adjust its polling scheme accordingly and transmit beacons when a node is expected to wake up, thus minimizing idle listening times.

In the previous schemes, the node wakeup times are known to the coordinator. Another approach has been adopted in Traffic-Aware Dynamic MAC (TAD-MAC) [88], in which the coordinator tries to learn the sleeping pattern of the nodes. TAD-MAC operation takes place in two phases. During the evolution phase, the coordinator polls the nodes at different times to see if they have packets to transmit, thus gathering statistics on their traffic load that are stored in a register bank. Different policies are employed for in-body (implantable) and on-body (wearable) nodes. In the first case, a star topology is considered and the register bank is maintained at the coordinator, whereas in the second case, a mesh topology is assumed, and nodes keep statistics for their neighbors traffic. The second operation phase begins after several cycles, when the protocol converges to a steady state and the beacon interval is adapted to the wakeup times of each node.

Finally, another way of coordinating transmissions in order to improve energy efficiency is by employing an additional low-power passive receiver at the nodes. This idea has been adopted in [89], where the nodes are equipped with two radio interfaces: (i) the main transceiver for the transmission of data; and (ii) a passive receiver with ultra-low power consumption, employed as a wakeup radio. During sleep mode, the nodes switch off their main radio interface and maintain only the secondary radio. The nodes exit sleep mode and activate the main transceiver only after receiving a wakeup signal by the coordinator in the secondary channel.

Synchronization mechanisms:*The implementation of all these energy-efficient schemes usually requires some synchronization between the coordinator and the nodes. A widely adopted synchronization mechanism is the transmission of timestamps within control packets sent by the coordinator (usually beacons or ACKs) [79,80,82,87]. Another approach is proposed in [86], where a predefined time interval, called wakeup fallback time, is employed. In case of communication failure, master and slave reattempt communication after this mutually known fallback time elapses. A more advances synchronization scheme called Adaptive Guard Band Algorithm (AGBA) with Drift Adjustment Factor (DAF) has been implemented within the TDMA-based MedMAC protocol [90]. The AGBA introduces two guard bands in the beginning and the end of each time slot. These guard bands have a variable duration which dynamically increases as time elapses from the previous synchronization point. The DAF is a correction factor employed for the calculation of the guard band duration, taking into account the estimated time offset between the clocks of the nodes and the coordinator. Finally, specific healthcare scenarios can offer unique opportunities for synchronization mechanisms. A typical example is H-MAC [91], a TDMA-based scheme that uses heartbeat rhythm to achieve synchronization. Exploiting the fact that many biosignals share the cardiac rhythm pattern, nodes placed on a single patient can process their sensory data to extract a common time reference, without the need of turning on their radio to receive beacons.*

4.4.2. QoS Provisioning Mechanisms

As mentioned in Section 3, healthcare applications have very heterogeneous traffic characteristics. Usually, most applications produce low-rate and periodic traffic, thus encouraging reservation-based scheduling solutions for uplink communication at the MAC layer. However, high traffic applications, such as video monitoring or aggregated data of given sensors (e.g., ECG, EMG, etc.)

must also be supported. Hence, MAC protocols should define different levels of priority and support advanced scheduling policies, often based on the cross-layer design paradigm.

Service differentiation and access priorities: *Priority-based scheduling based on service differentiation is one of the most widely employed mechanisms for QoS. The main idea lies in the assignment of different access priorities to traffic flows in order to achieve specific performance goals. Different criteria can be employed for the priority assignment, such as QoS requirements (e.g., latency constraints) or data criticality (e.g., emergency alarms versus normal data traffic).*

In the context of WBANs, a common approach is to distinguish between normal and emergency traffic and assign a higher level of access priority to the latter. This concept is adopted in the Urgency-based MAC (U-MAC) [92], a protocol based on the slotted Aloha mechanism employed in the IEEE 802.15.4. U-MAC classifies all traffic flows as either critical or non-critical. Critical data flows are allowed a higher number of transmission attempts, thus increasing their channel access probability. A different way to handle emergency data has been proposed in [86]. The idea is that whenever the master is not busy communicating with a slave that is awake, it sequentially sends polls to all associated slaves. If a slave node wants to communicate an alarm, it stays awake until a poll is received (otherwise, under normal traffic, the slave would wake up according to its sleep pattern). In [87], alarms are handled with the help of a secondary wakeup radio: the node informs the coordinator on the presence of emergency data by sending a control packet in the wakeup radio, and then the coordinator immediately polls the node in the main channel to retrieve the alarm message.

More advanced service differentiation schemes define more priority levels. For example, in BodyQoS [93], three classes of descending priority are defined: reserved downlink data from the coordinator to the nodes, reserved uplink data from the nodes to the coordinator and best-effort data. Reserved downlink data are scheduled first in each superframe, followed by the reserved uplink data, if there is still available bandwidth. Any remaining bandwidth is devoted to best-effort traffic. Adaptive resource scheduling that takes into consideration the channel quality is also provided. Two levels of scheduling are defined in [29]. At the first level, traffic flows are classified according to their criticality level (i.e., high, medium or low urgency). At the second level, each priority class is allocated a dynamically adjusted non-zero portion of the bandwidth according to the Generalized Processor Sharing principle, whereas Earliest Deadline First scheduling is adopted for traffic flows of the same class.

Cross-layer scheduling: *In addition to service differentiation, advanced scheduling algorithms can be designed by taking into account cross-layer information acquired from the interaction of the MAC with other protocol layers. Cross-layer parameters often include channel state information, provided by the PHY layer, or routing information from the network layer, among others.*

A simple channel-aware method, called Flipping algorithm, that can improve the performance of TDMA-based schemes, has been proposed in [94]. The solution applies to channels that can be modeled as a two-state Markov chain: good channel condition ensures successful data transmissions, whereas packet losses occur when the link quality is bad. The Flipping algorithm schedules the transmissions within a TDMA superframe based on the outcome of the previous superframe. All good links of the previous frame are scheduled first, but with reversed order, in order to exploit the most recently observed good quality links. On the other hand, all bad links are scheduled last in the new superframe, thus increasing the probability of having their channel quality changed to the best. BANMAC [95] is a more advanced cross-layer scheme that exploits the fact that, even though the channel quality of WBANs experiences many fluctuations (e.g., due to the movement of body parts), some regular patterns can still be observed. BANMAC is able to predict the channel fluctuations and schedule the transmissions of the nodes opportunistically when the channel conditions are more favorable. Another cross-layer approach is adopted in CICADA [81], where a joint MAC scheduling and routing scheme is employed to ensure ensure low-latency and network flexibility to topology changes.

Although QoS is the main objective of the scheduling mechanisms presented in this section, sometimes scheduling may also result to a reduction of energy consumption. In continuation, we present some schemes that include energy-related parameters such as power constraints (e.g., power limitations of implantable devices) or battery lifetime levels in the scheduling decisions.

In BSN-MAC [96], for example, four priority classes are defined to match four energy levels: (i) unconstrained, for the coordinator node that is assumed to have no energy constraints; (ii) constrained, for wearable sensor devices; (iii) highly constrained, for ingestible sensor devices; and (iv) extremely constrained, for implantable devices. The protocol also defines three levels of data criticality (low, normal and high). These parameters, along with the remaining battery lifetime level and the buffer occupancy, are combined to produce a cross-layer metric used to adjust the length of the IEEE 802.15.4 superframe with the objective of reducing latency while enhancing energy efficiency.

DQBAN takes into account the quality of the wireless link, the residual battery lifetime and the waiting time of the nodes in the transmission queue to implement a cross-layer, fuzzy-logic scheduler [83]. The main idea is that each node can affect the scheduling order by sending a specific request in the scheduling part of the frame. For example, a node may request to be given priority if its energy level becomes critical or to delay its transmission if bad channel quality is detected.

A cross-layer TDMA scheme that aims to increase the battery lifespan of the nodes without compromising reliability and QoS requirements, has been proposed in [97]. The main driver of the proposed scheme is based on battery discharge dynamics which dictate that battery lifetime can be increased if the battery remains idle for a certain amount of time. In order to maximize the idle time between transmissions, nodes may choose to accumulate several packets in their buffers. A decision on whether to transmit in a given time slot is made depending on cross-layer information regarding the battery recovery level, the buffer occupancy and the wireless channel state.

Finally, PEH-QoS [98] promotes energy-aware management and QoS provisioning in a WBAN with energy harvesting capabilities by proposing a three-module architecture that comprises: (i) the Power-EH Aware Management (PHAM), which ensures the optimal use of the harvested energy, by preventing transmission attempts by nodes with insufficient energy levels and encouraging the detection of all events by the sensor nodes (to ensure the detection of critical medical events); (ii) the Data Queue Aware Control (DQAC), which prevents the saturation of the data packet queue and maintains the clinical validity of the data by discarding outdated packets; (iii) the Packet Aggregator/Scheduling System (PASS) that maximizes throughput and energy efficiency by supporting the transmission of aggregated data packets.

4.4.3. Reliability

The reliable data transmission is another key issue addressed in the MAC layer. Since redundancy of devices is not a possibility, given the unique role played by each node in the WBAN, the MAC protocol must provide solutions to increase reliability. This can be achieved by supporting retransmission mechanisms or by increasing diversity through techniques such as cooperative network coding (NC).

Retransmission mechanisms: *A usual approach is to employ ACK packets to acknowledge the correct data reception (e.g., [82,83,87]). In case of erroneous reception, the data packet can either be discarded or a retransmission can be requested. In [80], for example, extra slots are reserved for packet retransmissions and their optimal number is calculated as a function of the packet error rate.*

Diversity techniques based on NC: *Cooperation between nodes is a key mechanism that can be used to enhance diversity. Exploiting the broadcast wireless nature, nodes in the vicinity of the transmitter (relays) can actively contribute to the communication by forwarding any overheard packets to the destination. Recently, cooperative NC techniques have been studied as a potential way to enhance reliability in WBANs.*

The work presented in [99] shows the potential throughput improvement under error-prone channels through the application of network coding, as a function of the number of redundancy packets, the employed relays and the number of sink nodes. More advanced schemes, such as cooperative diversity coding where both coded and uncoded data packets are transmitted, can yield ever better results [100]. A tree topology has been assumed in [101], where multiple nodes communicate with the coordinator through a small number of relays. The nodes send their uncoded data to the relays, which in turn forward a set of both coded and uncoded packets to the destination. The obtained results show that network coding can offer higher reliability with respect to traditional redundancy schemes where multiple retransmissions of uncoded packets take place. Similar conclusions are drawn in [102], where the coding decisions take into account the reliability requirements of different traffic flows.

Finally, a novel cloud-assisted MAC protocol, CLNC-MAC, that employs NC and guarantees the successful delivery of all generated data to the destination through a relay network, regardless of the channel conditions, has been proposed in [30]. The key concept behind CRNC-MAC is to employ cloud resources for the relay coordination, to ensure that all crucial information from the source is propagated without losses through the network (that may include multiple hops), thus enabling the successful decoding at the destination. This is achieved by employing a central coordinating entity, namely the cloud controller, which has the task of verifying the reception of all required information by the relays at a hop by hop basis, requesting retransmissions whenever necessary.

4.4.4. Interference and Coexistence Mechanisms

Interference-aware mechanisms aim to increase the robustness of WBANs to errors due to strong interference sources operating in the same license-free frequency bands, such as IEEE 802.11 or Bluetooth devices, often present in medical environments. Many existing works embrace the concept of cognition, by dynamically selecting the best channel for data transmission from a pool of available frequency bands, based on specific metrics, such as the channel quality or the interference levels. Coexistence with other wireless technologies that operate in the same frequency bands is also a key issue.

Interference-aware channel selection: *A cognitive radio approach for mHealth services is considered in [103]. Two spectrum sensing mechanisms are proposed to detect channels free of primary user activity. Between sensing periods, beacons are transmitted by the coordinator to distribute periodic bandwidth allocations to the nodes for normal traffic, whereas a small time interval is reserved immediately after the beacon transmission for the communication of emergency events.*

In [29], Multi-channel Quality-based MAC (MQ-MAC) is proposed in a scenario where the nodes placed on each patient form a WBAN cluster, and cluster coordinators form a mesh network to forward the aggregated patient data to a monitoring unit. In MQ-MAC, the cluster coordinators create and maintain a table with the link quality of all available channels. For each point-to-point link, the transmitter and the receiver agree on the most suitable channel for their communication and reserve it through a 3-way handshake on the dedicated control channel. This idea is taken one step further in the same work to form the Channel Quality Based Routing (CQBR) scheme, where the route with the best link quality is selected and enable simultaneous transmissions on different channels by neighboring nodes. Co-channel interference provoked by multiple co-located WBANs is also tackled by BANMAC [95], by a fully distributed channel allocation mechanism.

Coexistence mechanisms: *The Centralized BAN Access Scheme (CBAS), proposed in [104], tackles interference issues in a coexistence scenario with IEEE 802.11 devices. In CBAS the coordinator has cognitive radio capabilities and is able to scan the available spectrum continuously. When a suitable channel is found, the coordinator follows the IEEE 802.11 access rules to reserve the channel for the estimated time required for the data exchange by transmitting a legacy RTS packet. After that, it issues a wakeup command to the next scheduled node on a dedicated secondary control channel, thus waking up the node from its sleep mode and informing it on the data channel selection.*

Finally, a different interference-related challenge is explored in [105], where the proposed scheme aims to reduce the Electromagnetic Interference (EMI) produced by WBANs, which can lead to the malfunctioning of EMI-sensitive life-supporting medical devices (such as defibrillators) in a hospital environment. A context-aware solution is proposed, in which a coordinating device dynamically determines the transmission power limits for each communication in the WBAN, after consulting an inventory of the location and the EMI constraints of all electronic medical devices in a given setting. The nodes must content for access by transmitting an RTS in the dedicated control channel. The coordinator replies with a negative CTS if no feasible power scheme is available at that specific time due to either EMI constraints or predicted congestion conditions. Otherwise, a positive CTS with the maximum supported power allocation is transmitted, and the node enters a queuing system, waiting for its turn to transmit in the data channel.

The presented state-of-the-art approaches for the key MAC layer mechanisms, namely energy-efficient channel access, QoS provisioning, reliability, and interference and coexistence are summarized in Table 4.

5. END-TO-END SOLUTIONS FOR M2M COMMUNICATION

The goal of an mHealth application is to provide a bridge between the patient and the medical personnel. Hence, the M2M system must provide end-to-end connectivity, connecting the medical sensor devices via the M2M gateway to the Internet, and ultimately to the application server. This role is played by the access communication network that is usually based on WLAN/WMAN wireless technologies such as WiFi (IEEE 802.11), WiMAX (IEEE 802.16) and LTE/LTE-A. Since standardization efforts on M2M communications are still under way, there are many different approaches to achieve end-to-end connectivity.

Table 4. Overview of Medium Access Control (MAC) layer mechanisms for Wireless Body Area Networks (WBANs).

MAC Mechanism	Proposed Approach	Related Works
	Energy-efficient channel access	
Collision-free access	TDMA for periodic allocations	[79,80,90,91,94,97,103,104]
	Tree-based distributed scheduling	[81]
Hybrid access (contention-based/collision-free)	Contention phase for access request	[82]
	Distributed queuing with collision-free data transmission	[83]
	Energy-aware polling and probabilistic hybrid access	[84,85]
Preamble sampling, minimizing idle listening	Polling at predefined wake-up times	[86,87]
	Polling with dynamic learning of wake-up times	[88]
	Polling through secondary receiver	[89]
Synchronization	Timestamp through control packets	[79,80,82,83,87,88]
	Predefined wake-up fallback time	[86]
	Secondary channel	[89]
	Dynamic guard bands and clock drift correction	[90]
	Heartbeat-based time reference	[91]
	QoS Provisioning	
Access priorities	Two-level priorities (normal, alarms)	[86,87,92]
	Multiple priorities	[29,93]
Cross-layer scheduling	Channel-aware	[94,95]
	Routing-aware	[81]
	Energy-aware	[83,96–98]
	Reliability	
Retransmission mechanisms	ACK packets and retransmissions	[80,82,83,86,87,89,91,96]
Diversity with network coding (NC)	Transmission of NC packets	[99]
	Cooperative diversity coding	[100,101]
	Cooperative diversity coding with QoS	[102]
	Cloud-assisted NC	[30]
	Interference and coexistence mechanisms	
Interference-aware channel selection	Cognitive radio approach	[103]
	Channel selection based on link quality	[29,95]
Coexistence mechanisms	Coexistence with IEEE 802.11	[104]
	Coexistence in hospital environments	[105]

In this section, we examine end-to-end solutions for M2M communications. In particular, we first discuss theoretical works that focus on the high-level integration of WLAN/WMAN access communication technologies with WBANs. Then, we give some practical examples of end-to-end solutions by presenting testbed implementations for healthcare monitoring. Finally, we present some current research projects for mHealth applications.

5.1. Technology Integration for M2M Communications

The integration of access communication technologies with WBANs is a challenging but crucial issue. Some works identify the challenges to enable M2M communications over LTE and LTE-A, focusing on the air interface [106], scheduling mechanisms [107], and the support of low-cost devices [108]. The use of WiMAX-based broadband wireless access technology for mHealth applications, along with potential deployment scenarios and related radio resource management issues has been discussed in [109].

Apart from these high-level approaches, there are some works in the literature that study more specific M2M mHealth scenarios. For instance, [110] discusses the suitability of WiFi-based solutions within healthcare facilities such as single hospital units, as long as proper design, installation and validation methods are followed to meet the specific application requirements.

In [111], a two-tier network architecture is considered. At the lower tier, multiple sensor devices worn by a single patient form a WBAN, by employing the CSMA/CA access mode of IEEE 802.15.4. At the upper tier, multiple WBAN coordinators (corresponding to multiple patients) located within a specific area (e.g., a hospital ward), communicate with an Access Point through WiFi. End-to-end packet delay and access time has been modeled as a function of the number of coexisting WBANs. An extension of this work, in [112], has introduced service differentiation to prioritize high-rate data streams (e.g., EEG data) by providing them contention-free access, whereas CSMA/CA access is given to low-rate streams (e.g., ECG data).

In [113], the authors discuss the feasibility of employing a hybrid network based on WiFi and WiMAX technologies as the access communication network in an M2M system for mHealth services. The integration of the two technologies poses several challenges, mainly related to QoS provisioning, connection admission control, scheduling, and mobility management through seamless vertical handovers. The envisioned heterogeneous deployment scenario considers nodes with either single (WiFi or WiMAX) or dual radio interface and aims to provide wireless connectivity between different subnetworks, including WBANs, home care networks, mobile patients and networks of healthcare providers (such as intranets of hospitals, clinics, drugstores, etc.).

Another remote monitoring scheme that provides ubiquitous connectivity for mobile patients has been presented in [114]. In the proposed scheme, shown in Figure 5, a patient-attached monitoring device collects the WBAN data, classifies them as high-priority (e.g., critical data such as blood pressure, pulse rate and heart rate) or normal priority (e.g., ECG signal) and forwards them towards the healthcare provider through an heterogeneous WiFi/WiMAX access communication network. The access technology is selected depending on the patient's location, considering that WiFi hotspots cover only specific (mainly indoor) locations and WiMAX has a wider (outdoor) coverage. In addition, two types of connections are provided by the network operator: (i) low-cost reserved connections, allocated to patients for given amounts of time (e.g., weeks); and (ii) high-cost on-demand connections, employed when the available bandwidth for reserved connections is not enough to cover the traffic load. The authors approach this mHealth scenario from the service provider's side, who has to buy in advance a certain number of reserved connections from the network operator to serve a given number of patients. Stochastic programming techniques are used to determine the optimal number of reserved connections for each wireless technology in order to minimize the provider's cost.

As far as LTE-based solutions for healthcare applications are concerned, the relevant works in the literature are limited. The impact of 4G communication technologies on mHealth and the emerging challenges are discussed in [115]. In [116], a mapping between the QoS requirements of mHealth services and the existing service classes defined by 3GPP standards is proposed, aiming to provide guidelines for network operators. A cross-layer design for QoS of medical video streaming over mobile WiMAX (IEEE 802.16e) and the 3G High-Speed Packet Access (HSPA) cellular technology is proposed in [117], and a comparison between the performance of the two technologies is given, opening the road to further investigation on LTE-based mHealth applications.

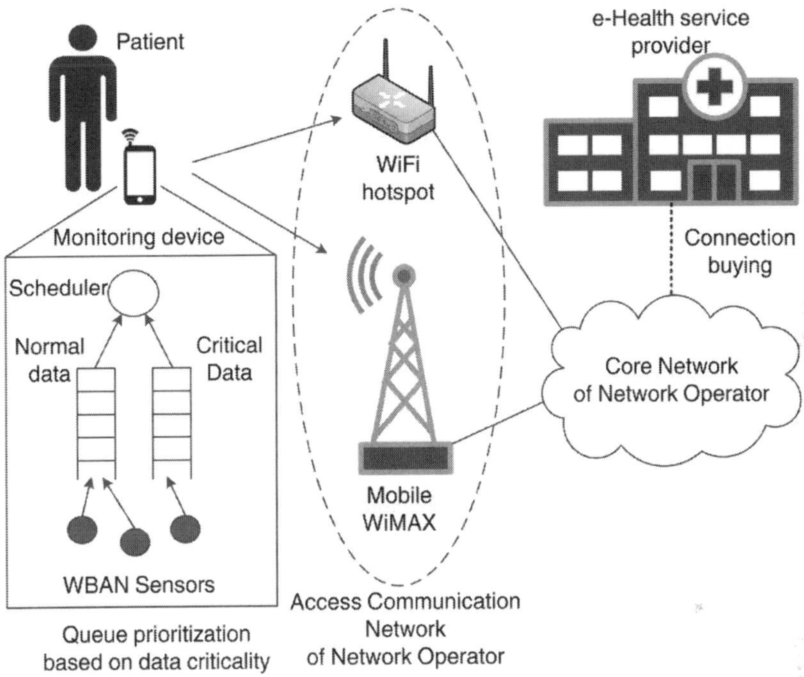

Figure 5. Architecture of remote patient monitoring system for WiFi/WiMAX heterogeneous scenario.

5.2. Testbed Implementation of M2M Solutions

Research on end-to-end connectivity is not limited to a theoretical-only level. During the last years, research efforts have been devoted to the actual implementation of M2M networks. In this section, we present the most representative examples of implemented end-to-end solutions for mHealth applications.

In [118], the authors introduce a new file format for the transfer of sensory data and implement a pilot testbed for an end-to-end patient monitoring application. Their contribution is twofold. First, they present an enhanced version of a standard protocol for communication among ECG devices, by proposing an adaptive data structure that can handle multiple vital signals, as well as data for positioning, allergies and demographic information on patients. The definition of a standardized data structure is an important step towards the integration of the medical data measured by the WBAN sensors with various mHealth information systems for monitoring or administrative purposes, belonging to hospitals, individual care-givers, home-care, etc. Second, they implement a testbed of an M2M healthcare application for the remote monitoring of patients suffering from heart problems. The patient is equipped with a WBAN formed by a number of wearable sensors, a Global Positioning System (GPS) device and a PDA. The PDA aggregates the sensory and geolocation data, as well as any additional information inserted manually by the patient, and plays the role of the M2M gateway. On the one hand, it employs Bluetooth technology to communicate with the WBAN nodes and, on the other hand, it has mobile ADSL capabilities to forward the data to the remote server located at a hospital facility. To the other end of the system, a portable data acquisition system is considered, consisting of a medical monitor device, a GPS and a laptop with Internet connectivity. Finally, a software application has been developed for the processing and visualization of the data retrieved by the healthcare provider. The pilot testing of the proposed solution on real patients has revealed some very interesting conclusions. From the doctors' perspective, the use of the M2M mHealth system has been an overall positive experience, facilitating the patient monitoring and the collection of data. The patients, on the other hand, have given a more neutral evaluation. Even though they have generally been satisfied by the experience, they have shown more concerns on the wearability of the sensors, the user friendliness of the software application and the data collection process.

A two-tier architecture is considered in [119] to implement a remote monitoring application for patients with Chronic Obstructive Pulmonary Disease (COPD). Bluetooth is used as the WBAN technology for the communication between the sensors and the coordinating node (e.g., a PDA). Apart from the Bluetooth network interface, the coordinating node can support two additional long-range wireless technologies for Internet connectivity with a remote medical server: cellular General Packet Radio Service (GPRS) and WiFi. The authors perform an interesting experiment by measuring system performance metrics of these two upper-tier technologies. The study clearly shows that GPRS and WLAN have complementary power and delay profiles: the energy consumption of GPRS is low but high delays may be observed, whereas WiFi has higher energy cost but lower delays. Based on the observed results, the authors provide guidelines for the design of an adaptive protocol that switches between the two long-range technologies depending on the scenario: (i) WiFi is the recommended technology, when available, especially in emergency situations due to the low data latency. In these cases, the GPRS interface should remain on but at an idle state, employed only to receive incoming calls if needed; (ii) if the WiFi connection is not available, the WiFi network interface should be switched off completely and GPRS should be employed for communication.

A prototype M2M solution built on Android mobile devices is presented in [120]. At the patient's side, wearable sensors are employed to measure physiological data such as ECG and oxygen saturation. The sensor nodes are connected to M2M nodes, which are, in turn, wirelessly connected to the Internet through the M2M gateway. The sensor network employs the IEEE 802.15.4 protocol along with the IPv6 addressing mechanism. At the other end of the M2M system, a server PC collects and processes the sensor data, which are ultimately made available to the medical personnel through a user-friendly Android application. By combining Ipv6 techniques over mobile communication technology, this testbed implementation has provided a practical, scalable and flexible way to make a large amount of biomedical signals accessible online.

In some works, ambient sensor networks for environmental monitoring are employed in conjunction with WBANs, in order to provide additional information on the patient's whereabouts, such as temperature, humidity and light conditions. Along this line, a three-tier network architecture, depicted in Figure 6, is proposed in [121], for the remote monitoring of elderly or chronic patients in their residence. The lower tier consists of two systems: (i) a patient-worn fabric belt, which integrates the medical sensors and is equipped with a Bluetooth transceiver; and (ii) the ambient wireless sensors that form a ZigBee network and are deployed in the patient's surroundings (e.g., in the patient's home or in a nursing house). In the middle tier, an ad hoc network of powerful mobile computing devices (e.g., laptops, PDAs, etc.) gathers the medical and ambient sensory data and forwards them to the higher tier. The middle-tier devices must have multiple network interfaces: Bluetooth and ZigBee to communicate with the lower tier and WLAN or cellular capabilities for connection with the higher layer. Finally, the higher tier is structured on the Internet and includes the application databases and servers that are accessed by the healthcare providers. The study involves a real implementation of the proposed architecture and tackles several security issues that arise along the three tiers. The proposed framework offers a flexible and secure solution for the monitoring of multiple patients that can be applied to different scenarios, including home, hospital and nursing home environments.

Sensor networks can also be employed for patient localization purposes. In [122], the authors propose a system architecture based on two independent subsystems for the monitoring and location tracking of patients within hospital environments, as shown in Figure 7. The healthcare monitoring subsystem consists of smart shirts with integrated medical sensors, each equipped with a wireless IEEE 802.15.4 module. The location subsystem has two components: (i) a deployment of wireless IEEE 802.15.4 nodes that are installed in known locations within the hospital infrastructure and broadcast periodic beacon frames; and (ii) IEEE 802.15.4 end devices, held by the patients, that collect signal strength information from the received beacons. Both subsystems transmit their respective data (i.e., medical sensory data and signal strength information) to a gateway through an IEEE 802.15.4-based ad hoc distribution network. The gateway has wired Internet connectivity and forwards the data to the management server and the monitoring mHealth application. The proposed system has been tested with success in a hospital, achieving high reliability, sufficient battery lifetime of the sensors and real-time data reconstruction. In terms of usability, the obtained feedback of the medical personnel has been taken into account to improve the software interface.

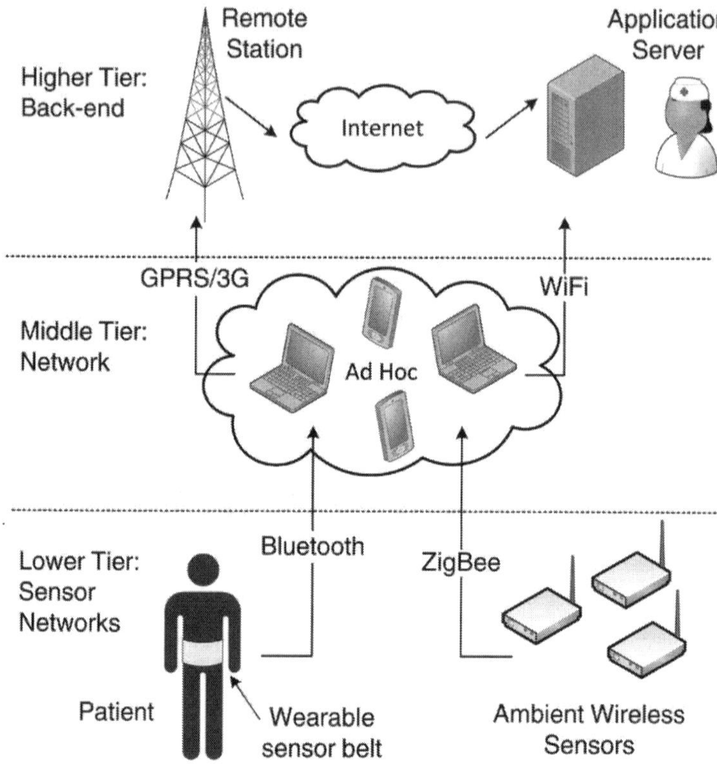

Figure 6. Example of three-tier network architecture.

5.3. Projects for End-to-End Connectivity

The aim of this section is to present an overview of the most recent and relevant research funded projects for mHealth applications.

HEALTH@HOME (Health at Home) [123] aims to provide an end-to-end solution for the remote monitoring of cardiovascular and respiratory patient parameters. The data are continuously gathered through an automatic processing system and are accessed by the responsible medical personnel. A typical client/server architecture is adopted, where the client side is a residential gateway located at the patient's home, able to collect data from the biomedical sensors through wireless Bluetooth links. The most significant measured signals are ECG, SpO2, weight, blood pressure, chest impedance, respiration and body posture. The measured data are sent through the gateway to a server located at the health service facilities that is integrated with the Hospital Information System. The gateway communicates with the server through ADSL as the primary transmission channel, or mobile broadband (i.e., GSM/GPRS/UMTS) as the secondary (backup) data channel. Alarms are sent by Short Message Service

(SMS) directly to the physicians, the patients' relatives and their caregivers. The HTTPS protocol addresses the security issues in the communication between the gateway and the server through a certificate validation process. The proposed solution has been tested on 30 patients during monitoring periods of at least one month and has received positive feedback by both patients and medical personnel, as a reliable, user-friendly means of remote control and management of acute conditions.

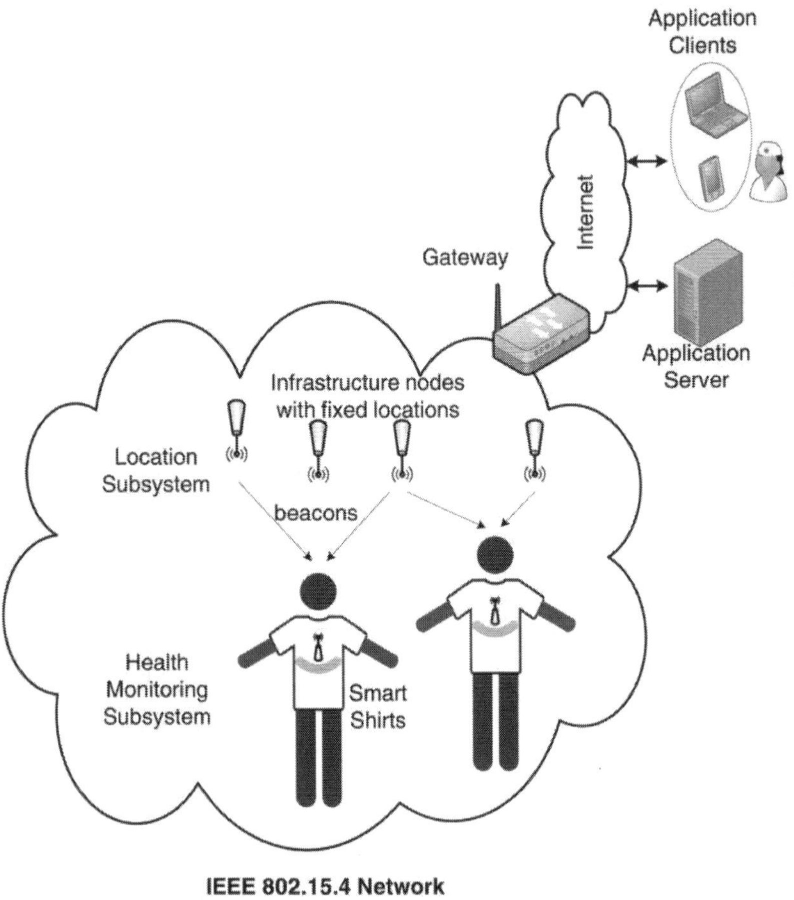

Figure 7. Example of the monitoring and location tracking mHealth system architecture.

HELP: Home-based Empowered Living for Parkinson Disease Patients [124] targets at designing a health monitoring system able to control disease progression and to mitigate Parkinson Disease (PD) symptoms, thus improving the quality of life of affected elderly people. Although it provides an end to end solution that employs M2M communication for monitoring patients with PD, its aim is to design a control system for a subcutaneous infusion pump that administers the exact required drug dose according to the patients' level of activity without focusing in communication issues. This system is composed of the following components: (i) an intra-oral electronic drug delivery device with miniaturized, non-invasive and removable design; (ii) an external pump that delivers higher amounts of drug; (iii) a WBAN to gather information on the user environment to detect blockades; (iv) a telecommunication and services infrastructure to analyze and transfer data exchanged between the user and the automated system; and (v) a remote care unit for patient supervision.

CAALYX-MV: Complete Ambient Assisted Living Experiment—Market Validation [125] goal is to provide an end to end solution that is focused on improving the elder's quality of life. The proposed solution is composed of: (i) a home system capable of monitoring and controlling social and health status of elder people and providing them with some tools and services to support their daily activities; (ii) a roaming system that comprises a smart textile shirt able to measure specific vital signs, detect falls and communicate emergencies; and (iii) a care system for the monitoring of individuals by family, caretakers and health services. All sensors in the WBAN are wearable, measure different parameters such as motion, blood pressure and heart rate, and communicate using Bluetooth links with a mobile phone. The sensory data are sent through standard low-cost networking equipment to a GPS-enabled smart phone (3G/UMTS) that runs a completely autonomous software application. The application continuously analyzes sensor data in order to identify problematic conditions and promptly alert the care system. The proposed system will be validated through 3 pilot programs that will test the usability and acceptability of the system by the users (both patients and caregivers) and will evaluate the reliability and detection accuracy of health problems in the monitored patients.

Help4Mood [126] aims at developing an end-to-end system to help the recovery of people with major depression. The system is designed to be used together with other forms of therapy, such as self-help, counseling or medication. The main components include: (i) a personal monitoring system to keep track of important behavior aspects, comprised of sensors for both user activity and sleep monitoring; (ii) an interactive virtual agent asking patients about their health and well-being and providing a portal to trusted health information; and (iii) a decision support system handling the virtual agent to allow its customization to the individual needs of the person with depression.

The sensor devices communicate by using a proprietary low-power RF network protocol named SimpliciTI [127] over Bluetooth. To increase energy efficiency and reliability, the system adopts the idea of cooperation between nodes, achieved through the slight modification of the MAC protocol.

WSN4QoL: Wireless Sensor Networks for Quality of Life [128] is a project focused on wireless communication technologies for mHealth applications. The main objectives of WSN4QoL are: (i) to provide a protocol stack architecture, which can accommodate a variety of protocols, algorithms and sensor devices for healthcare applications; (ii) to develop reliable, energy-efficient, interference-robust communication protocols and algorithms; (iii) to develop distributed localization protocols that meet the constraints imposed by WBANs in health care scenarios; and (iv) to propose effective and efficient security solutions for the proposed communication protocols. The proposed protocols and algorithms will be integrated in healthcare commercial devices, in order to evaluate the performance improvements in realistic environments.

To conclude, Table 5 provides a comparison among the main characteristics of the projects presented here. Summarizing, the presented works and projects study different aspects of M2M systems for healthcare delivery, ranging from solving technical communication problems to the implementation of close to market solutions. Despite their differences, several common goals can be identified, that can serve as guidelines for the design of successful end-to-end mHealth applications: (i) miniaturization and enhanced wearability of the sensor devices, to provide unobstructive monitoring that will not interfere with normal life activities of the patients; (ii) reliable two-way communication protocols, that guarantee the prompt and successful delivery of data from the medical sensors to the medical personnel, as well as the reception of medical feedback at the patient; (iii) accurate detection of emergency situations to ensure timely medical intervention in life threatening events. In addition, it is important to maintain a low probability of false positive alarms, to avoid unnecessary hospitalizations and interventions; (iv) advanced security mechanisms to guarantee confidentiality and privacy of the medical data; and (v) user friendly and easy to learn application interfaces, to ensure the successful adoption of the mHealth solutions, given that patients are often elderly people not familiar with the use of technology. Furthermore, the visualization of the monitored data must be done in a clear and helpful way for both the patient and the healthcare providers.

Table 5. Summary of existing projects on mHealth applications.

Project	Application	M2M Area Net.	M2M Com. Net.	Security
HEALTH@HOME	Cardiovascular disease	Bluetooth	ADSL	✓
HELP	Parkinson disease	Bluetooth	–	✗
CAALYX-MV	Independent living	Bluetooth	3G UMTS / WiFi	✓
Help4Mood	Depression management	Bluetooth/SimpliciTI	ADSL	✓
WSN4QoL	Remote disease management	Cloud-based distribution network	IEEE 802.15.6	✓

6. CONCLUSIONS

This paper has provided a systematic study on M2M systems for mHealth applications from a wireless communication perspective. We have, then, focused on end-to-end connectivity in M2M systems. After discussing the integration challenges between diverse communication technologies, we have highlighted different design approaches for end-to-end connectivity through examples of practical testbed implementations for healthcare services. Finally, a list of recent research projects in the context of mHealth has been given, with emphasis on the different technical solutions adopted in each project. Based on the high-level ETSI M2M system architecture, we have first focused on the device domain, by presenting an overview of the recently proposed PHY layer technologies and MAC layer design approaches that tackle the specific challenges of mHealth applications for WBANs.

Concluding, despite the considerable amount of work conducted in the rapidly growing area of M2M communications for mHealth, there are still many open issues to be addressed. Apart from the specific challenges in each aspect of the mHealth systems, efforts must be concentrated on standardization activities, that will enable the market exploitation of the scientific contributions in this field by paving the road for the development of interoperable M2M mHealth solutions.

ACKNOWLEDGMENTS

This work has been funded by the research projects WSN4QoL (286047), SGR(2014 SGR 1551) and GEOCOM (TEC2011-27723-C02-01).

AUTHOR CONTRIBUTIONS

All the authors have contributed in various degrees to ensure the quality of this work. Elli Kartsakli was responsible for the paper's organization and provided the M2M network architecture, the design challenges for mHealth applications and the MAC protocols overview. Aris S. Lalos focused on the WBAN channel modeling and the PHY layer technologies. Angelos Antonopoulos provided to the end-to-end technology integration. Stefano Tennina was responsible for summarizing the key testbeds and projects. Finally, Marco di Renzo, Luis Alonso and Christos Verikoukis supervised the work and provided valuable feedback on PHY technologies, MAC layer design and M2M systems, respectively.

CONFLICTS OF INTEREST

The authors declare no conflicts of interest.

REFERENCES

1. OECD Report. Machine-to-Machine Communications: Connecting Billions of Devices. Available online: (accessed on 23 September 2014).

2. Chang, K.; Soong, A.; Tseng, M.; Xiang, Z. Global Wireless Machine-to-Machine Standardization. IEEE Internet Comput. **2011**, 15, 64–69.

3. OneM2M Global Partnership Project, 2012. Available online: http:// www.onem2m.org/ (accessed on 23 September 2014).

4. Research and Markets: Machine-to-Machine (M2M) Communication in Healthcare 2010-20: Reviews the Major Drivers and Barriers for Growth of M2M. Available online: (accessed on 23 September 2014).

5. European Coordination Committee of the Radiological, Electromedical and Healthcare IT Industry (COCIR). Available online: http:// www. cocir.org/site/fileadmin/Publications_2011/telemedicine_toolkit_link2.pdf (accessed on 23 September 2014).

6. ETSI. Machine to Machine Communications (M2M): Use Cases of M2M Applications for eHealth. Available online: http://www. etsi. org/deliver/etsi_tr/102700_102799/102732/01.01.01_60/tr_102732v010101 p.pdf (accessed on 23 September 2014).

7. Ullah, S.; Kjan, P.; Ullah, N.; Saleem, S.; Higgins, H.; Kwak, K.S. A Review of Wireless Body Area Networks for Medical Applications. Int. J. Commun. Netw. Syst. Sci. **2009**, 2, 797–803.

8. Chakraborty, C.; Gupta, B.; Ghosh, S.K. A Review on Telemedicine-Based WBAN Framework for Patient Monitoring. Telemed. e-Health **2013**, 19, 619–626.

9. Lalos, A.S.; Alonso, L.; Verikoukis, C. Model based compressed sensing reconstruction algorithms for ECG telemonitoring in {WBANs}. Digit. Signal Process. **2014**.

10. Wu, G.; Talwar, S.; Johnsson, K.; Himayat, N.; Johnson, K. M2M: From mobile to embedded internet. IEEE Commun. Mag. **2011**, 49, 36–43. [Google Scholar]

11. Chen, K.C. Machine-to-Machine Communications for Healthcare. J. Comput. Sci. Eng. **2012**, 6, 119–126.

12. Shen, X. Emerging technologies for e-healthcare [Editor's Note]. IEEE Netw. **2012**, 26, 2–3. [Google Scholar]

13. Aragues, A.; Escayola, J.; Martinez, I.; del Valle, P.; Munoz, P.; Trigo, J.; Garcia, J. Trends and challenges of the emerging technologies toward interoperability and standardization in e-health communications. IEEE Commun. Mag. **2011**, 49, 182–188.

14. Ko, J.; Lu, C.; Srivastava, M.; Stankovic, J.; Terzis, A.; Welsh, M. Wireless Sensor Networks for Healthcare. IEEE Proc. **2010**, 98, 1947–1960.

15. Chen, M.; Gonzalez, S.; Vasilakos, A.; Cao, H.; Leung, V.C. Body Area Networks: A Survey. Mob. Netw. Appl. **2011**, 16, 171–193.

16. Latré, B.; Braem, B.; Moerman, I.; Blondia, C.; Demeester, P. A survey on wireless body area networks. Wirel. Netw. **2011**, 17, 1–18.

17. Alemdar, H.; Ersoy, C. Wireless sensor networks for healthcare: A survey. Comput. Netw. **2010**, 54, 2688–2710.

18. Patel, M.; Wang, J. Applications, challenges, and prospective in emerging body area networking technologies. IEEE Wirel. Commun. **2010**, 17, 80–88. [Google Scholar]

19. ETSI. Machine to Machine Communications (M2M): Functional Architecture. Available online: http://www.etsi.org/deliver/ etsi_ts/102600_102699/102690/01.01.01_60/ts_102690v010101p.pdf (accessed on 23 September 2014).

20. LAN/MAN Standards Committee of the IEEE Computer Society. In IEEE Standard for Local and Metropolitan Area Networks—Part 15.6: Wireless Body Area Networks; IEEE Std 802.15.6-2012: Piscataway, NJ, USA, 2012.

21. Movassaghi, S.; Abolhasan, M.; Lipman, J.; Smith, D.; Jamalipour, A. Wireless Body Area Networks: A Survey. IEEE Commun. Surv. Tutor. **2014**, 16, 1658–1686.

22. Cavallari, R.; Martelli, F.; Rosini, R.; Buratti, C.; Verdone, R. A Survey on Wireless Body Area Networks: Technologies and Design Challenges. IEEE Commun. Surv. Tutor. **2014**, 16, 1635–1657.

23. Dishongh, T.J.; McGrath, M. Wireless Sensor Networks for Healthcare Applications; Artech House: Norwood, MA, USA, 2009.

24. Jedermann, R.; PÃűtsch, T.; Lloyd, C. Communication techniques and challenges for wireless food quality monitoring. Philos. Trans. R. Soc. A Math. Phys. Eng. Sci. **2014**, 372, 1471–2962.

25. Tsouri, G.R.; Prieto, A.; Argade, N. On Increasing Network Lifetime in Body Area Networks Using Global Routing with Energy Consumption Balancing. Sensors **2012**, 12, 13088–13108.

26. Smith, D.; Miniutti, D.; Hanlen, L. Characterization of the Body-Area Propagation Channel for Monitoring a Subject Sleeping. IEEE Trans. Antennas Propag. **2011**, 59, 4388–4392.

27. Shi, X.; Medard, M.; Lucani, D. Whether and where to code in the wireless packet erasure relay channel. IEEE J. Sel. Areas Commun. **2013**, 31, 1379–1389.

28. Lalos, A.S.; Kartsakli, E.; Antonopoulos, A.; Tennina, S.; Renzo, M.D.; Alonso, L.; Verikoukis, C. Cooperative Compressed Sensing schemes for Telemonitoring of Vital Signals in WBANs. IEEE Globecom **2014**. accepted.

29. Chen, B.; Pompili, D. Transmission of patient vital signs using wireless body area networks. Mob. Netw. Appl. **2011**, 16, 663–682.

30. Kartsakli, E.; Antonopoulos, A.; Alonso, L.; Verikoukis, C. A Cloud-Assisted Random Linear Network Coding Medium Access Control Protocol for Healthcare Applications. Sensors **2014**, 14, 4806–4830.

31. Gandhi, V. M2M Security—The Need for a Holistic Approach. Available online: http://www.frost.com/sublib/display-market-insight-top.do?id=240308296 (accessed on 23 September 2014).

32. Ryckaert, J.; de Doncker, P.; Meys, R.; de le Hoye, A.; Donnay, S. Channel model for wireless communication around human body. Electron. Lett. **2004**, 40, 543–544.

33. Zasowski, T.; Meyer, G.; Althaus, F.; Wittneben, A. Propagation effects in UWB body area networks. Proceedings of the IEEE International Conference on Ultra-Wideband (ICU 2005), Zurich, Switzerland, 5–8 September 2005.

34. Reusens, E.; Joseph, W.; Latre, B.; Braem, B.; Vermeeren, G.; Tanghe, E.; Martens, L.; Moerman, I.; Blondia, C. Characterization of On-Body Communication Channel and Energy Efficient Topology Design for Wireless Body Area Networks. IEEE Trans. Inf. Technol. Biomed. **2009**, 13, 933–945.

35. Zasowski, T.; Althaus, F.; Stager, M.; Wittneben, A.; Troster, G. UWB for noninvasive wireless body area networks: Channel measurements and results. Proceedings of the IEEE Conference on Ultra Wideband Systems and Technologies (UWBST 2003), Reston, VA, USA, 16–19 November 2003; pp. 285–289.

36. Hashemi, H. The indoor radio propagation channel. IEEE Proc. **1993**, 81, 943–968.

37. Fort, A.; Desset, C.; Ryckaert, J.; de Doncker, P.; van Biesen, L.; Wambacq, P. Characterization of the ultra wideband body area propagation channel. Proceedings of the IEEE International Conference on Ultra-Wideband (ICU 2005), Zurich, Switzerland, 5–8 September 2005.

38. Nadeau, P.M. Multi-Channel Ultra-Low-Power Receiver Architecture for Body Area Networks. Master's Thesis; Massachusetts Institute of

Technology, Department of Electrical Engineering and Computer Science, 2011. Available online: http://dspace.mit.edu/handle/1721.1/66476 (accessed on 23 September 2014).

39. Roelens, L.; van den Bulcke, S.; Joseph, W.; Vermeeren, G.; Martens, L. Path loss model for wireless narrowband communication above flat phantom. Electron. Lett. **2006**, 42, 10–11.

40. Fort, A.; Desset, C.; Ryckaert, J.; de Doncker, P.; van Biesen, L.; Donnay, S. Ultra wide-band body area channel model. Proceedings of the IEEE International Conference on Communications (ICC 2005), Seoul, Korea, 16–20 May 2005; pp. 2840–2844.

41. Taparugssanagorn, A.; Rabbachin, A.; Hämäläinen, M.; Saloranta, J.; Iinatti, J. A Review of Channel Modelling for Wireless Body Area Network in Wireless Medical Communications. Proceedings of the 11th International Symposium on Wireless Personal Multimedia Communications, Lapland, Finland, 8–11 September 2008.

42. Yazdandoost, K.Y.; Kohno, R. Body Implanted Medical Device Communications. IEICE Trans. **2009**, E92.B, 410–417.

43. Sani, A.; Alomainy, A.; Hao, Y. Numerical Characterization and Link Budget Evaluation of Wireless Implants Considering Different Digital Human Phantoms. IEEE Trans. Microw. Theory Tech. **2009**, 57, 2605–2613.

44. Cho, N.; Roh, T.; Bae, J.; Yoo, H.J. A Planar MICS Band Antenna Combined With a Body Channel Communication Electrode for Body Sensor Network. IEEE Trans. Microw. Theory Tech. **2009**, 57, 2515–2522.

45. Gupta, S.; Lalwani, S.; Prakash, Y.; Elsharawy, E.; Schwiebert, L. Towards a propagation model for wireless biomedical applications. Proceedings of the IEEE International Conference on Communications (ICC 2003), Anchorage, AK, USA, 11–15 May 2003; pp. 1993–1997.

46. Tang, Q.; Tummala, N.; Gupta, S.; Schwiebert, L. Communication scheduling to minimize thermal effects of implanted biosensor networks in homogeneous tissue. IEEE Trans. Biomed. Eng. **2005**, 52, 1285–1294.

47. Johansson, A. Wave-propagation from medical implants-influence of body shape on radiation pattern. Proceedings of the 2nd Joint Engineering 24th Annual Conference in Medicine and Biology and the Annual Fall Meeting of the Biomedical Engineering Society EMBS/BMES Conference, Houston, TX, USA, 23–26 October 2002; Volume 2, pp. 1409–1410.

48. Wegmueller, M.; Kuhn, A.; Froehlich, J.; Oberle, M.; Felber, N.; Kuster, N.; Fichtner, W. An Attempt to Model the Human Body as a Communication Channel. IEEE Trans. Biomed. Eng. **2007**, 54, 1851–1857.

49. Hachisuka, K.; Terauchi, Y.; Kishi, Y.; Hirota, T.; Sasaki, K.; Hosaka, H.; Ito, K. Simplified circuit modeling and fabrication of intrabody communication devices. Proceedings of the 13th International Conference on Transducers, Solid-State Sensors, Actuators and Microsystems (TRANSDUCERS 2005), Seoul, Korea, 5–9 June 2005; pp. 461–464.

50. Liu, X.; Yang, X.; Wang, Y.; Wang, L. Comprehensive measurements on body channel characteristics of Human Body Communication. Proceedings of the International Conference on Microwave and Millimeter Wave Technology (ICMMT 2010), Chengdu, China, 8–11 May 2010; pp. 1098–1101.

51. Zimmerman, T.G. Personal area networks: Near-field intrabody communication. IBM Syst. J. **1996**, 35, 609–617.

52. Oh, J.Y.; Kim, J.H.; Lee, H.S.; Kim, J.Y. PSSK Modulation Scheme for High-Data Rate Implantable Medical Devices. IEEE Trans. Inf. Technol. Biomed. **2010**, 14, 634–640.

53. Bohorquez, J.; Chandrakasan, A.; Dawson, J. A 350 μW CMOS MSK Transmitter and 400 μW OOK Super-Regenerative Receiver for Medical Implant Communications. IEEE J. Solid-State Circuits **2009**, 44, 1248–1259.

54. Abouei, J.; Brown, J.; Plataniotis, K.; Pasupathy, S. Energy Efficiency and Reliability in Wireless Biomedical Implant Systems. IEEE Trans. Inf. Technol. Biomed. **2011**, 15, 456–466.

55. Huan-Bang, Li.; Kamya Yekeh Yazdandoost, B.Z. Wireless Body Area Network; The River Publishers Series in Information Science and Technology; River Publishers: Aalborg, Denmark, 2010.

56. Alomainy, A.; Hao, Y.; Hu, X.; Parini, C.; Hall, P. UWB on-body radio propagation and system modelling for wireless body-centric networks. IEE Proc. Commun. **2006**, 153, 107–114.

57. Niemela, V.; Rabbachin, A.; Taparugssanagorn, A.; Hamalainen, M.; Iinatti, J. A comparison of UWB WBAN receivers in different measured hospital environments. Proceedings of the 3rd International Symposium on Applied Sciences on Biomedical and Communication Technologies (ISABEL 2010), Rome, Italy, 7–10 November 2010; pp. 1–5.

58. Di Renzo, M.; Buehrer, R.; Torres, J. Pulse Shape Distortion and Ranging Accuracy in UWB-Based Body Area Networks for Full-Body Motion Capture and Gait Analysis. Proceedings of the IEEE Global Telecommunications Conference (GLOBECOM '07), Washington, DC, USA, 26–30 November 2007; pp. 3775–3780.

59. Dotlić, I.; Miura, R. Low data-rate ultra-wideband impulse radio communications for wireless medical body area networks: Highly-reliable noncoherent system design. Proceedings of the 4th International

Symposium on Applied Sciences on Biomedical and Communication Technologies (ISABEL 2011), New York, NY, USA, 26–29 October 2011; pp. 1–5.

60. Zasowski, T.; Wittneben, A. Performance of UWB Receivers with Partial CSI Using a Simple Body Area Network Channel Model. IEEE J. Sel. Areas Commun. **2009**, 27, 17–26. [Google Scholar]

61. Gupta, B.; Cianca, E.; Ruggieri, M.; Prasad, R. A novel FM-UWB system for vital sign monitoring and its comparison with IR-UWB. Proceedings of the 2nd International Symposium on Applied Sciences on Biomedical and Communication Technologies (ISABEL 2009), Bratislava, Slovakia, 24–27 November 2009; pp. 1–4.

62. Gupta, B.; Valente, D.; Cianca, E.; Prasad, R. FM-UWB for radar and communications in medical applications. Proceedings of the First International Symposium on Applied Sciences on Biomedical and Communication Technologies (ISABEL 2008), Aalborg, Denmark, 25–28 October 2008; pp. 1–5.

63. Viittala, H.; Hamalainen, M.; Iinatti, J. Impact of difference in WBAN channel models on UWB system performance. Proceedings of the IEEE 11th International Symposium on Spread Spectrum Techniques and Applications (ISITA 2010), Taichung, Taiwan, 17–20 October 2010; pp. 175–180.

64. Abbasi, Q.H.; Sani, A.; Alomainy, A.; Hao, Y. On-Body Radio Channel Characterization and System-Level Modeling for Multiband OFDM Ultra-Wideband Body-Centric Wireless Network. IEEE Trans. Microw. Theory Tech. **2010**, 58, 3485–3492.

65. Batra, A.; Balakrishnan, J.; Aiello, G.; Foerster, J.; Dabak, A. Design of a multiband OFDM system for realistic UWB channel environments. IEEE Trans. Microw. Theory Tech. **2004**, 52, 2123–2138.

66. Shi, K.; Zhou, Y.; Kelleci, B.; Fischer, T.; Serpedin, E.; Karsilayan, A. Impacts of Narrowband Interference on OFDM-UWB Receivers: Analysis and Mitigation. IEEE Trans. Signal Process. **2007**, 55, 1118–1128.

67. Islam, S.M.; Kwak, K.S. A Comprehensive Study of Channel Estimation for WBAN-based Healthcare Systems: Feasibility of Using Multiband UWB. J. Med. Syst. **2012**, 36, 1553–1567.

68. Cho, N.; Yoo, J.; Song, S.J.; Lee, J.; Jeon, S.; Yoo, H.J. The Human Body Characteristics as a Signal Transmission Medium for Intrabody Communication. IEEE Trans. Microw. Theory Tech. **2007**, 55, 1080–1086.

69. Schenk, T.; Mazloum, N.; Tan, L.; Rutten, P. Experimental characterization of the body-coupled communications channel. Proceedings of the IEEE International Symposium on Wireless Communication Systems (ISWCS 2008), Reykjavik, Iceland, 21–24 October 2008; pp. 234–239.

70. Cho, N.; Lee, J.; Yan, L.; Bae, J.; Kim, S.; Yoo, H.J. A 60 kb/s-to-10 Mb/s 0.37 nJ/b Adaptive-Frequency-Hopping Transceiver for Body-Area Network. Proceedings of the IEEE International Solid-State Circuits Conference Digest of Technical Papers (ISSCC 2008), San Francisco, CA, USA, 3–7 February 2008; pp. 132–602.

71. Fazzi, A.; Ouzounov, S.; van den Homberg, J. A 2.75 mW wideband correlation-based transceiver for body-coupled communication. Proceedings of the IEEE International Solid-State Circuits Conference Digest of Technical Papers (ISSCC 2009), San Francisco, CA, USA, 8–12 February 2009; pp. 204–205.

72. Ryckaert, J.; Desset, C.; Fort, A.; Badaroglu, M.; de Heyn, V.; Wambacq, P.; van der Plas, G.; Donnay, S.; van Poucke, B.; Gyselinckx, B. Ultra-wide-band transmitter for low-power wireless body area networks: Design and evaluation. IEEE Trans. Circuits Syst. I Regul. Pap. **2005**, 52, 2515–2525.

73. Zhang, L.-H.; Liu, X.-Z.; Huang, J.; Wang, L. Baseband system for human body channel communication. Proceedings of the 3rd International Conference on Biomedical Engineering and Informatics (BMEI 2010), Yantai, China, 16–18 October 2010; pp. 778–781.

74. Demirkol, I.; Ersoy, C.; Alagoz, F. MAC protocols for wireless sensor networks: A survey. IEEE Commun. Mag. **2006**, 44, 115–121.

75. Kredo, K., II; Mohapatra, P. Medium Access Control in wireless sensor networks. Comput. Netw. **2007**, 51, 961–994.

76. Bachir, A.; Dohler, M.; Watteyne, T.; Leung, K. MAC Essentials for Wireless Sensor Networks. IEEE Commun. Surv. Tutor. **2010**, 12, 222–248.

77. Ullah, S.; Shen, B.; Riazul Islam, S.M.; Khan, P.; Saleem, S.; Kwak, K.S. A Study of Medium Access Control Protocols for Wireless Body Area Networks. Sensors **2010**, 10, 128–145.

78. Gopalan, S.; Kim, D.H.; Nah, J.W.; Park, J.T. A survey on power-efficient MAC protocols for wireless body area networks. Proceedings of the 3rd IEEE International Conference on Broadband Network and Multimedia Technology (IC-BNMT 2010), Beijing, China, 26–28 October 2010; pp. 1230–1234.

79. Milenkoviċ, A.; Otto, C.; Jovanov, E. Wireless sensor networks for personal health monitoring: Issues and an implementation. Comput. Commun. **2006**, 29, 2521–2533.

80. Marinkovic, S.; Popovici, E.; Spagnol, C.; Faul, S.; Marnane, W. Energy-Efficient Low Duty Cycle MAC Protocol for Wireless Body Area Networks. IEEE Trans. Inf. Technol. Biomed. **2009**, 13, 915–925.

81. Latré, B.; Braem, B.; Moerman, I.; Blondia, C.; Reusens, E.; Joseph, W.; Demeester, P. A Low-delay Protocol for Multihop Wireless Body Area Networks. Proceedings of the 4th Annual International Conference on Mobile and Ubiquitous Systems: Networking Services (MobiQuitous 2007), Philadelphia, PA, USA, 6–10 August 2007; pp. 1–8.

82. Fang, G.; Dutkiewicz, E. BodyMAC: Energy efficient TDMA-based MAC protocol for Wireless Body Area Networks. Proceedings of the 9th International Symposium on Communications and Information Technology (ISCIT 2009), Incheon, Korea, 28–30 September 2009; pp. 1455–1459.

83. Otal, B.; Alonso, L.; Verikoukis, C. Highly reliable energy-saving MAC for wireless body sensor networks in healthcare systems. IEEE J. Sel. Areas Commun. **2009**, 27, 553–565.

84. Ibarra, E.; Antonopoulos, A.; Kartsakli, E.; Rodrigues, J.; Verikoukis, C. Energy Harvesting Aware Hybrid MAC Protocol for WBANs. Proceedings of the IEEE 15th International Conference on IEEE e-Health Networking, Applications Services (Healthcom 2013), Lisbon, Portugal, 9–12 October 2013; pp. 120–124.

85. Ibarra, E.; Antonopoulos, A.; Kartsakli, E.; Verikoukis, C. HEH-BMAC: Hybrid Polling MAC Protocol for Wireless Body Area Networks Operated by Human Energy Harvesting. Springer Telecommun. Syst. J. **2012**. accepted.

86. Omeni, O.; Wong, A.; Burdett, A.; Toumazou, C. Energy Efficient Medium Access Protocol for Wireless Medical Body Area Sensor Networks. IEEE Trans. Biomed. Circuits Syst. **2008**, 2, 251–259.

87. Ullah, S.; Kwak, K. An Ultra Low-power and Traffic-adaptive Medium Access Control Protocol for Wireless Body Area Network. J. Med. Syst. **2012**, 36, 1021–1030.

88. Alam, M.; Berder, O.; Menard, D.; Sentieys, O. TAD-MAC: Traffic-Aware Dynamic MAC Protocol for Wireless Body Area Sensor Networks. IEEE J. Emerg. Sel. Top. Circuits Syst. **2012**, 2, 109–119.

89. Zhang, X.; Jiang, H.; Chen, X.; Zhang, L.; Wang, Z. An energy efficient implementation of on-demand MAC protocol in medical Wireless Body Sensor Networks. Proceedings of the IEEE International Symposium on Circuits and Systems (ISCAS 2009), Taiper, Taiwan, 24–27 May 2009; pp. 3094–3097.

90. Timmons, N.; Scanlon, W. Improving the ultra-low-power performance of IEEE 802.15.6 by adaptive synchronisation. IET Wirel. Sens. Syst. **2011**, 1, 161–170.

91. Li, H.; Tan, J. Heartbeat-Driven Medium-Access Control for Body Sensor Networks. IEEE Trans. Inf. Technol. Biomed. **2010**.

92. Ali, K.; Sarker, J.; Mouftah, H. Urgency-Based MAC Protocol for Wireless Sensor Body Area Networks. Proceedings of the IEEE International Conference on Communications Workshops (ICC 2010), Cape Town, South Africa, 23–27 May 2010; pp. 1–6.

93. Zhou, G.; Lu, J.; Wan, C.Y.; Yarvis, M.; Stankovic, J. BodyQoS: Adaptive and Radio-Agnostic QoS for Body Sensor Networks. Proceedings of the IEEE 27th Conference on Computer Communications (INFOCOM 2008), Phoenix, AZ, USA, 13–18 April 2008; pp. 565–573.

94. Tselishchev, Y. Designing a Medium Access Control protocol for Body Area Networks. Proceedings of the IEEE International Symposium on a World of Wireless, Mobile and Multimedia Networks (WoWMoM 2011), Lucca, Italy, 20–24 June 2011; pp. 1–3.

95. Prabh, K.; Royo, F.; Tennina, S.; Olivares, T. BANMAC: An Opportunistic MAC Protocol for Reliable Communications in Body Area Networks. Proceedings of the IEEE 8th International Conference on Distributed Computing in Sensor Systems (DCOSS 2012), Hangzhou, China, 16–18 May 2012; pp. 166–175.

96. Li, H.; Tan, J. An Ultra-low-power Medium Access Control Protocol for Body Sensor Network. Proceedings of the 27th Annual International Conference of the Engineering in Medicine and Biology Society (IEEE-EMBS 2005), Shanghai, China, 1–4 September 2005; pp. 2451–2454.

97. Su, H.; Zhang, X. Battery-dynamics driven TDMA MAC protocols for wireless body-area monitoring networks in healthcare applications. IEEE J. Sel. Areas Commun. **2009**, 27, 424–434.

98. Ibarra, E.; Antonopoulos, A.; Kartsakli, E.; Rodrigues, J.; Verikoukis, C. Joint Power-QoS Control Scheme for Energy Harvesting Body Sensor Nodes. Proceedings of the IEEE International Conference on Communications (ICC 2014), Sydney, Australia, 10–14 June 2014.

99. Arrobo, G.; Gitlin, R. Improving the reliability of wireless body area networks. Proceedings of the Annual International Conference of the IEEE Engineering in Medicine and Biology Society (EMBC 2011), Boston, MA, USA, 30 August–3 September 2011; pp. 2192–2195.

100. Arrobo, G.; Gitlin, R. New approaches to reliable wireless body area networks. Proceedings of the IEEE International Conference on Microwaves, Communications, Antennas and Electronics Systems (COMCAS 2011), Tel Aviv, Israel, 7–9 November 2011; pp. 1–6.

101. Marinkovic, S.; Popovici, E. Network Coding for Efficient Error Recovery in Wireless Sensor Networks for Medical Applications. Proceedings of the 1st International Conference on Emerging Network Intelligence (EMERGING 2009), Sliema, Malta, 11–16 October 2009; pp. 15–20.

102. Taparugssanagorn, A.; Ono, F.; Kohno, R. Network Coding for Non-Invasive Wireless Body Area Networks. Proceedings of the IEEE 21st International Symposium on Personal, Indoor and Mobile Radio Communications Workshops (PIMRC Workshops 2010), Istanbul, Turkey, 26–29 September 2010; pp. 134–138.

103. Feng, S.; Liang, Z.; Zhao, D. Providing telemedicine services in an infrastructure-based cognitive radio network. IEEE Wirel. Commun. **2010**, 17, 96–103.

104. Torabi, N.; Leung, V. Robust license-free Body Area Network access for reliable public m-health services. Proceedings of the 13th IEEE International Conference on e-Health Networking Applications and Services (Healthcom 2011), Columbia, MO, USA, 13–15 June 2011; pp. 312–319.

105. Phunchongharn, P.; Niyato, D.; Hossain, E.; Camorlinga, S. An EMI-Aware Prioritized Wireless Access Scheme for e-Health Applications in Hospital Environments. IEEE Trans. Inf. Technol. Biomed. **2010**, 14, 1247–1258.

106. Lien, S.Y.; Chen, K.C.; Lin, Y. Toward ubiquitous massive accesses in 3GPP machine-to-machine communications. IEEE Commun. Mag. **2011**, 49, 66–74.

107. Gotsis, A.; Lioumpas, A.; Alexiou, A. M2M Scheduling over LTE: Challenges and New Perspectives. IEEE Veh. Technol. Mag. **2012**, 7, 34–39.

108. Beale, M. Future challenges in efficiently supporting M2M in the LTE standards. Proceedings of the IEEE Wireless Communications and Networking Conference Workshops (WCNCW 2012), Paris, France, 1 April 2012; pp. 186–190.

109. Niyato, D.; Hossain, E.; Diamond, J. IEEE 802.16/WiMAX-based broadband wireless access and its application for telemedicine/e-health services. IEEE Wirel. Commun. **2007**, 14, 72–83.

110. Baker, S.; Hoglund, D. Medical-Grade, Mission-Critical Wireless Networks [Designing an Enterprise Mobility Solution in the Healthcare Environment]. IEEE Eng. Med. Biol. Mag. **2008**, 27, 86–95.

111. Misic, J.; Misic, V. Bridging between IEEE 802.15.4 and IEEE 802.11b networks for multiparameter healthcare sensing. IEEE J. Sel. Areas Commun. **2009**, 27, 435–449.

112. Misic, J.; Misic, V. Bridge performance in a multitier wireless network for healthcare monitoring. IEEE Wirel. Commun. **2010**, 17, 90–95. [Google Scholar]

113. Zhang, Y.; Ansari, N.; Tsunoda, H. Wireless telemedicine services over integrated IEEE 802.11/WLAN and IEEE 802.16/WiMAX networks. IEEE Wirel.Commun. **2010**, 17, 30–36.

114. Niyato, D.; Hossain, E.; Camorlinga, S. Remote patient monitoring service using heterogeneous wireless access networks: Architecture and optimization. IEEE J. Sel. Areas Commun. **2009**, 27, 412–423.

115. Istepanaian, R.S.H.; Zhang, Y.T. Guest Editorial Introduction to the Special Section: 4 G Health—The Long-Term Evolution of m-Health. IEEE Trans. Inf. Technol. Biomed. **2012**, 16, 1–5.

116. Skorin-Kapov, L.; Matijasevic, M. Analysis of QoS requirements for e-health services and mapping to evolved packet system QoS classes. Int. J. Telemed. Appl. **2010**, 9, 1–18.

117. Alinejad, A.; Philip, N.; Istepanian, R. Cross-Layer Ultrasound Video Streaming Over Mobile WiMAX and HSUPA Networks. IEEE Trans. Inf. Technol. Biomed. **2012**, 16, 31–39.

118. Mandellos, G.J.; Koukias, M.N.; Styliadis, I.S.; Lymberopoulos, D.K. e-SCP-ECG+ protocol: An expansion on SCP-ECG protocol for health telemonitoring—Pilot implementation. Int. J. Telemed. Appl. **2010**, 2010, 1–17.

119. Wac, K.; Bargh, M.; van Beijnum, B.J.; Bults, R.; Pawar, P.; Peddemors, A. Power- and delay-awareness of health telemonitoring services: The mobihealth system case study. IEEE J. Sel. Areas Commun. **2009**, 27, 525–536.

120. Jung, S.J.; Myllyla, R.; Chung, W.Y. Wireless Machine-to-Machine Healthcare Solution Using Android Mobile Devices in Global Networks. IEEE Sens. J. **2013**, 13, 1419–1424.

121. Huang, Y.; Hsieh, M.; Chao, H.; Hung, S.; Park, J. Pervasive, secure access to a hierarchical sensor-based healthcare monitoring architecture in wireless heterogeneous networks. IEEE J. Sel. Areas Commun. **2009**, 27, 400–411.

122. López, G.; Custodio, V.; Moreno, J. LOBIN: E-Textile and Wireless-Sensor-Network-Based Platform for Healthcare Monitoring in Future Hospital Environments. IEEE Trans. Inf. Technol. Biomed. **2010**, 14, 1446–1458.

123. Health@Home Project. Available online: http://www.aal-europe.eu/projects/healthhome/ (accessed on 23 September 2014).

124. Home-based Empowered living for Parkinson's Diseade Patients Project (HELP). Available online: http://www.aal-europe.eu/projects/help/ (accessed on 23 September 2014).

125. CAALYX-MV Project. Available online: http://www.caalyx-mv.eu/project (accessed on 23 September 2014).

126. Help4Mood Project. Available online: http://www.help4mood.info/site/default.aspx (accessed on 23 September 2014).

127. SimpliciTI Protocol Stack. Available online: http://www.ti.com/tool/simpliciti (accessed on 23 September 2014).

128. WSN4QoL Project. Available online: http://www.wsn4qol.eu/ (accessed on 23 September 2014).

CHAPTER 4

A Preamble-Based Broadcasting Technique for Wireless Ad Hoc Sensor Networks

Arun Kumar, Kai-Juan Wong

Emerging Research Lab, School of Computer Engineering,
Nanyang Technological University, Singapore

ABSTRACT

Broadcasting is a fundamental operation in any wireless networks, more so in wireless ad hoc sensor networks, where each sensor node has limited transmission range and battery power. Although broad-casting in wireless ad hoc sensor networks has many advantages but it can cause serious problems like—broadcast storm, which could cause a lot of contention, redundant, retransmission, collision and most importantly, drain immense amount of energy from limited battery powered sensor nodes. In this work, our objective is to reduce the number of retransmission and energy consumption of sensor nodes by using the duty cycle property of wireless ad hoc sensor networks. We propose a preamble-based broadcasting technique for wireless ad hoc sensor networks. We show that in dense wireless ad hoc sensor networks a small size preamble can give maximum network-wide data dissemination rather than using the large preamble, which will only consume immense amount of energy during packet reception.

KEYWORDS

Broadcast, Flooding, Wireless Ad Hoc Sensor Network

1. INTRODUCTION

Wireless ad hoc sensor networks are a set of wireless sensor nodes (mobile or static) which communicate with each other without any pre-existing routing infrastructure for communication, and mostly rely on their neighbor nodes to relay their data up to the destination node instead of directly communicating with the destination node or sink. These sensor nodes have several resource constraints such as limited memory, battery power, and signal processing, computational and communication capabilities. These tiny nodes are mainly deployed in remote areas and work unattended for many years, making battery life of the sensor node a major area of concern.

A large-scale wireless ad hoc sensor network can consist of thousands of sensor nodes and are helpful in many situations [1], such as disaster relief missions, battlefield communication facilities, infrastructure protection and scientific exploration [1-3]. Wireless data transmission consumes more energy than data processing in a sensor node and the situation become worse in dense wireless networks where broadcasting of data is needed. Even a single bit transmission consumes the same energy as that needed for processing thousand operations in a typical sensor node.

Wireless ad hoc sensor networks have broadcast as one of its most fundamental service. Broadcast provides maximum message propagation across the whole network and serves high-level operations, making it critical to the overall network design. Although, broadcasting in wireless ad hoc sensor networks have many advantages, it can cause serious problems like broadcast storm problems [4], which could cause a lot of contention, redundant retransmission, collision and most importantly waste immense amount of energy. A possible solution is that the sensor nodes alternate between active and dormant states, which help them to conserve energy and extend the network lifetime. Simple broadcasting in this environment, where a node goes to active and dormant states periodically can be very challenging if there are not enough nodes in the active state and when the source wants to send data.

In this paper, we propose a preamble based broadcasting technique for wireless ad hoc sensor networks, while considering the challenges raised by the broadcast storm problems in such networks. The proposed technique is implemented and studied thoroughly, taking mobile node in the wireless ad hoc sensor networks. A probability-based technique is also implemented and studied to compare with our technique. The set of performances matrics chosen for fair comparison are reachability, throughput, average number of duplicate packets and energy consumption.

The rest of the paper is configured as follows. Section 2 gives a background of broadcasting techniques for wireless networks. In Section 3, we present our proposed preamble-based broadcasting technique. The simulation platform used, the parameters used during the simulation and the matrics used for comparison are given in Section 4. Section 5 gives a thorough discussion of the results. Section 6 concludes the paper and gives suggestions for future work.

2. BACKGROUND

There are many proposed approaches for broadcasting in which, flooding is one of the earliest broadcasting mechanisms in wired and wireless networks. In flooding, each node rebroadcasts the packets when received for the first time, already received packets just dropped. Though flooding is simple, it consumes most of the network resources as a large number of duplicate packets flooded repeatedly. This can leads to serious redundancy, contention and collision in any wireless network and is referred as the "broadcast storm problem" [4].

The authors in [4] have classified broadcasting schemes into probabilistic, counter-based, distance based, location-based and cluster-based. In the probabilistic scheme, a host will rebroadcast the message with the maximum probability $p = 1$, if it receives for first time; otherwise every time the probability would be decreased. In the counter-based scheme, a host will drop the packet if the number of duplicates received crosses a threshold. In the distance-based scheme, a host will rebroadcast the packet only if the sender and receiver distance is larger than a threshold. The location-based scheme rebroadcast the message if the additional coverage due to this broadcast is bigger than a threshold. Finally, in the clusterbased scheme, clusters are created using efficient cluster selection algorithms and the cluster head and gateways do the rebroadcast. An alternate classification of broadcasting techniques found in [5] has four categories: simple flooding, probability-based, area-based, and neighbor knowledge-based schemes.

Several duty cycle aware broadcast protocols have been proposed [6-9] in recent years for wireless sensor networks. Duty cycle aware broadcast [6] shows that conventional broadcast protocol cannot cover the whole network in an acceptable timeframe and the performance degrades in low duty cycle environment. The protocol tried to balance between latency and efficiency with coverage guarantees by remodeling the broadcast problem for a low duty cycle environment.

In Opportunistic flooding [7] the decision to forward a packet is based on probabilistic forwarding decisions at the sender side, considering the delay distribution of next-hop nodes. A new node joining the network shares its schedule with the entire neighborhood and the process of sharing the schedule is referred to as low duty cycle rendezvous. To reduce redundancy in transmission and flooding delay, only opportunistically early packets are forwarded using links outside the energy optimal tree. A forwarder selection method to overcome the hidden terminal problem and a link-quality based back off method to resolve simultaneous forwarding operations are proposed by the author. Opportunistic flooding considers the working schedules of sensor nodes as fixed. The algorithm assumes that the network is locally synchronized using MAC layer time stamping technique, which makes the algorithm complex, as synchronization itself is a complex problem in big area network.

Asynchronous Duty-cycle Broadcasting (ADB) [8] is integrated with the MAC layer and uses the information available at this layer, unlike traditional multi-hop broadcast which operates above the MAC layer. ADB optimizes the broadcast progress in the network at the level—of transmission from the node to each neighbor individually. In ADB, a sender first transmits unicast packet to a node when it is in active state and through the acknowledgement, it learns if the neighbor has been reached through the broadcast. It is integrated with the RI-MAC [8] to announce its wake up with the beacon packets. ADB can efficiently perform in unicast but the performance in broadcast traffic scenarios is not thoroughly studied.

The solution to this problem can be made simpler by making use of a tree to flood packets. But this solution can be very fragile [9,10] because a single node failure (parent node) can prevent all its sub trees nodes from receiving messages, even though the network is fully connected. These tree solutions can be energy efficient at the cost of long delays, as the packets are always forwarded via a single route.

Unfortunately, more or less all the above mentioned protocols are concerned mostly about the quality of service in wireless ad hoc sensor network and do not consider the challenges raised by broadcasting such as redundant transmission, collision and contention.

In this paper, we have considered asynchronous duty cycling nodes during the simulation of our proposed broadcasting scheme. We have used a small preamble of variable length to overcome the vast energy consumption of sensor nodes and significantly reduce the number of retransmissions in wireless ad hoc sensor networks.

3. PREAMBLE-BASED BROADCASTING

Wireless ad hoc sensor networks have an important property that many sensor nodes alternate between active and dormant states, which help them to conserve energy and extend the network lifetime. In a dense area networks there is a greater chance of a node being in an active state to listen to the channel at a particular time, in contrast to a sparse area network. Simple flooding in dense area network, where more nodes are active at a time near the relay node will only cause energy consumption rather than covering additional areas in the network, resulting in unnecessary retransmission of packets and collisions.

We have used small preambles in dense wireless ad hoc sensor networks before the data packet. This small preamble before the data packet in the dense area network will be listened by a subset of nodes present in the radio range of a relay node or sender node. The subset of sensor nodes that received the data packet then rebroadcasts the packet using the same preamble length. This practice of sending a small preamble before the data packet reduces the number of rebroadcasts as well as the overall energy consumption in the network, as only a subset of nodes will listen to the preamble and receive the data packet. Each sensor node rebroadcast the packet only after a random delay to make the timing of rebroadcasting differentiated.

In a highly dense area network, a very small preamble can give maximum data dissemination in contrast to a comparatively large preamble required in a sparse area network where there are lesser chances that a node is active in the neighborhood of the relaying sensor node or the sender node. This will save most of the rebroadcasts and conserve maximum energy in dense area networks when compared to sparse networks, where small preamble cannot guarantee minimum number of node for network wide data dissemination.

Figure 1 compares simple and preamble based flooding. Figure 1(a) shows the effect of simple flooding in a wireless ad hoc network where every sensor node is involved in flooding and number of retransmission is large. Arrows are pointed both sides to show that a node is forwarding a packet to its neighbor node as well as getting the packet from its neighbor node to whom it forwarded the packet before. Figure 1(b) shows broadcasting using preamble, based technique in duty cycle networks, where a subset of sensor nodes receives a packet every time a node broadcasts it, thus providing network wide coverage with minimum retransmission of packets.

The relevance and benefits of using a small preamble before broadcast data packets to get network wide dissemination can be justified in following steps:

- Sensor nodes in the wireless ad hoc sensor networks are tiny devices with several constraints, such as limited radio range (20 - 30 m typically), memory, battery power, computation and communication capabilities. These devices are mainly deployed in remote areas and work non-stop 24 × 7 unattended for many years. As a result the main focus is on saving the battery power of the sensor node.

- Broadcasting method such as simple flooding can cause severe energy consumption on already battery power limited nodes.

- Redundant rebroadcast of message is undesirable in both dense and sparse wireless ad-hoc sensor networks. Redundant rebroadcast cause immense amount of energy wastage and at the same time is not of much help in distributing the message throughout the network.

- Broadcast packets in wireless ad hoc sensor networks are generally used for route discovery or for periodically updating the routing table. In this case, when broadcasting is needed periodically, simple flooding can consume most of the network resources, even before the actual data packet reaches to its destination (sink).

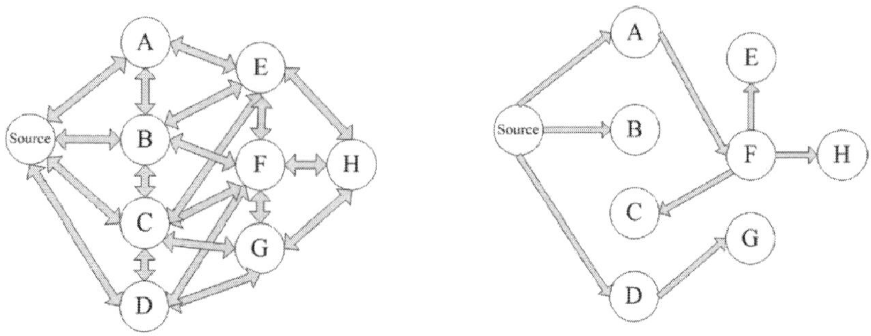

Figure 1. Different flooding structures. (a) Original flooding (b) Flooding using preamble technique in duty cycle network.

4. SIMULATION AND ANALYSIS

We have implemented our protocols using the QualNet 5.0 simulator. QualNet can be used to optimize new and existing models, analyze the performance of networks, perform cross-layer optimization of wired or wireless network stacks or design a network in QualNet Developer to perform capacity, performance, and scalability analyses. We have modified several layers of QualNet to implement BMAC and our proposed protocol.

4.1. Application Layer Implementation

The Application Layer of QualNet 5.0 does not provide broadcast based traffic-generating protocol, it provides several unicast and multicast based traffic generating protocol such as CBR, VBR, MCBR, CELLU-LARABSTRACT-APP and GSM.

We have implemented our own broadcast traffic generating protocol on QualNet 5.0. This enables the nodes present in the network to broadcast data packet network-wide. A node broadcasts the data packet just after receiving it from any other node, without any delay. Every broadcast packet is 32 byte in size and contains a sequence number as well as a timestamp, which contains the time the packet was sent by the other node.

4.2. MAC Layer Implementation

QualNet 5.0 MAC library provides several MAC Layer protocols, for both wireless and wired category and provides the flexibility of running different MAC protocols (wired and wireless mix) on a node, if a node has multiple interfaces. Some examples of the MAC Layer protocols available in qualnet are as ALOHA, CSMA, TDMA, GENERICMAC, MAC802.3, MAC802.15.4 and MACDOT11.

Although, QualNet 5.0 has a variety of MAC Layer protocols in its library, it lacks any Energy Aware MAC Layer protocol such as S-MAC, D-MAC, BMAC, X-MAC, RI-MAC and others. We have tried to fill this gap by implementing the most popular energy aware MAC layer protocol, BMAC in QualNet 5.0. To implement BMAC, the physical layer of QualNet is modified to simulate sleep and wakeup states of radio.

4.3. Physical Layer Implementation

The Physical Layer of QualNet 5.0 does not model the sleep and waking up states of the radios, which is assumed to be either in the transmission or receiving modes. The physical layer model is modified to allow the radios to go on sleep mode, when nothing is on the channel after a predefined time or after receiving the data packet. The channel is sensed periodically and if anything is detected on channel, the node will wake up the radio. Switching from sleep to idle takes significant time, which is considered as the waking up time and the state called as waking up state. Adding the sleep state in the physical layer will help the nodes in energy conservation by allowing them to control when to switch ON and OFF the radio. The physical layer of QualNet is modeled to be in one of the Sleep/Waking-up/Idle/Transmit/Receive states. Successful implementation of these states is mportant for the correct implementation of a duty cycling MAC protocol in QualNet Simulator.

4.4. Broadcasting in Qualnet

The simulation is a collection of network nodes, each with its own protocol stack parameters and statistics accessible from the "Node" structure. There are two types of events in QualNet: packet events and timer events.

The two types of messages/events that represent a broadcast event are discussed below. These are data structures that hold information on broadcast event type and associated data.

1) Packet events (data or control)—simulate exchange of data packets between layers or between nodes.

Timer events—simulates time-outs, allows broadcast protocol to schedule events in a future time; internal to a protocol. Timer events are set and received within a protocol and they do not travel through the protocol stack.

QualNet uses a layered architecture, similar to that of the TCP/IP network protocol stack. Within that architectture, data moves between adjacent layers. QualNet protocol stack consists of, from top to bottom, the Application, Transport, Network, Link (MAC) and Physical Layers. Adjacent layers in the protocol stack communicate via well-defined APIs shown in Table 1, and generally, layer communication occurs only between adjacent layers, as shown in Figure 2.

4.5. Simulation Parameters

4.5.1. Parameters for Preamble-Based Broadcasting

As stated earlier, we have simulated our work on QualNet 5.0. Due to the several constraints on wireless sensor nodes, we have fixed some parameters used in simulation, such as the radio range (20.023 m), data rate (250 kbps) and mobility of nodes as 2 m/s. Certain parameters are determined based on the CC2420 transceiver [12], such as transmit power consumption of radio (57.4 mW), receiver power consumption (62.1 mW), and power consumption during sleep mode (1.41 mW).

For the simulations, node 1 is always chosen as source node, generating the broadcast packet (1 packet/sec) and all nodes are mobile including the source node. Preamble length ranges from 4 ms to 21 ms. Some important parameters used during simulation are given in the Table 2.

Table 1. Message-related API functions.

MESSAGE_Alloc( )	Allocate a message and provide it with standard event, layer, protocol info
MESSAGE_InfoAlloc( )	Allocate additional user-specified space for optional information about the event
MESSAGE_PacketAlloc( )	Allocate space for the packet within the message
MESSAGE_Send( )	Send the message as an event to the specified layer and protocol
MESSAGE_Free()	Free the message, once it has reached its final destination

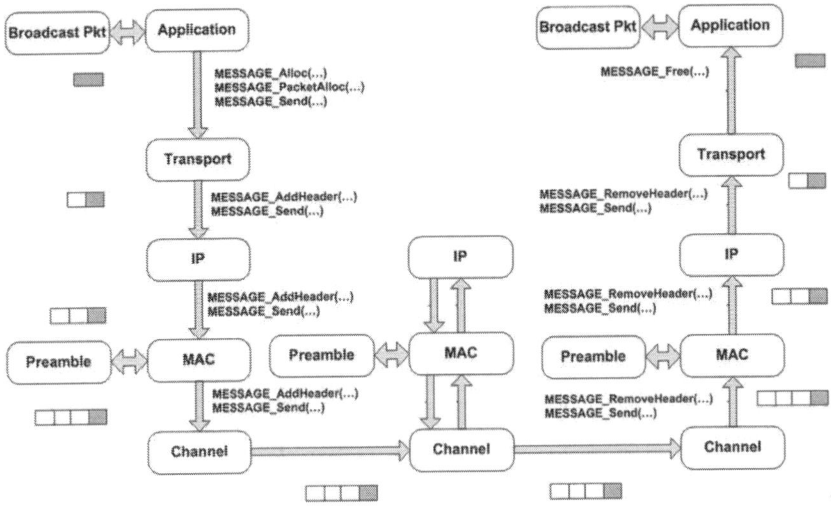

Figure 2. Broadcast packet protocol stack in qualnet.

4.5.2. Parameters for Probability-Based Broadcasting

The technique, probability based technique for broadcasting is also simulated using QualNet 5.0. Several parameter used during simulation are same as preamble based techniques such as the radio range (20.023 m), data rate (250 kbps), mobility of nodes (2 m/s) and the power consumption. Nodes in the network are not duty cycled and CSMA is used as MAC layer protocol.

Broadcast packets are sent using different probabilities, ranging from 0.2 to 1. Node 1 is always chosen as source node generating the broadcast packet (1 packet/sec) and all nodes are mobile including the source node. Some important parameters used during simulation are given in the Table 3.

A random delay of a few microseconds is taken on each node before rebroadcasting the packet, to make the timing of rebroadcasting differentiated in both the broadcasting technique.

4.6. Performance Metrics

The set of performance metrics used for analysis are as follows:

- REachability (RE): We define the reachability as the number of sensor nodes receiving the broadcast message divided by the total number sensor nodes that are reachable, directly or indirectly, from the source node.

- Throughput: We define the throughput as the average number of unique packet successfully received by a node over the total number of unique packets send by the source node.
- Average Duplicate Packet: We define the average duplicate packet as the average number of duplicate packet received by a node during the simulation time.
- Energy Consumption: Energy consumption for a particular network density is calculated as the average energy consumption of a node during the simulation time.

Table 2. Parameters used in simulation.

Application Layer	Broadcast
MAC Layer	BMAC
Path Loss Model	Two-Ray
Bandwidth	250 kbps
Payload Size	32 bytes
Terrain	50 × 50, 100 × 100, 200 × 200, 300 × 300, 400 × 400, 500 × 500
Node Placement	Uniform
Radio Range	20.023 m
Node Numbers	250
Mobility	2m/sec.
Wakeup Interval	20 ms
Simulation Time	500 sec.
Iterations	10

5. RESULTS AND DISCUSSION

Network Sizes shown in figures are 1 × 1, 2 × 2, 4 × 4, 6 × 6, 8 × 8, 10 × 10 units, where a unit represents 50 meter in length.

Table 3. Parameters used in simulation.

Application Layer	Broadcast
MAC Layer	CSMA
Path Loss Model	Two-Ray
Bandwidth	250 kbps
Payload Size	32 bytes
Terrain	50 × 50, 100 × 100, 200 × 200, 300 × 300, 400 × 400, 500 × 500
Node Placement	Uniform
Radio Range	20.023 m
Node Numbers	250
Mobility	2 m/sec.
Simulation Time	500 sec.
Iterations	10

5.1. Reachability

Figure 3 and Figure 4 show the percentage nodes reached versus network size, comparing the preamble-based broadcasting with the probability-based broadcasting. We observe that in dense area network broadcasting a packet with a small preamble or with a less probability is sufficient to disseminate the broadcast packet throughout the network. In contrast, as the network becomes sparse, a broadcast packet needs to be sent with a longer preamble or with a high probability. We observe that there is no significant difference in our preamble based broadcast as compared with the probability-based broadcast in terms of reachability. Our scheme provides higher reachability in dense area network and is a better solution in term of reachability in dense area wireless ad hoc sensors networks.

5.2. Throughput

Figure 5 and Figure 6 show the percentage throughput versus network size, comparing preamble-based broadcasting with probability-based broadcasting. We observe that in a very dense area network probability-based broadcasting fails to achieve high throughput due to the high collisions of packets. It is observed that in dense area networks preamble-based scheme achieves high throughput even if not sent with a long preamble. But when the network becomes sparser probability-based scheme achieves high throughput because the nodes are at all times active in the network. The abrupt drop in throughput in sparse area networks is due to minimum reachability. The high throughput in the probabilitybased schemes comes at a cost of high-energy consumption and duplicates packets. It is observe that our preamble-based scheme provides a better throughput while consuming less energy.

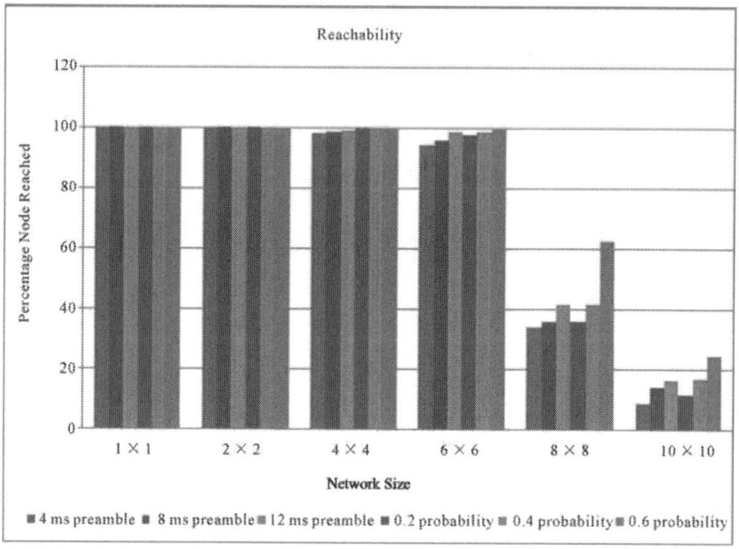

Figure 3. Percentage node reached vs network size.

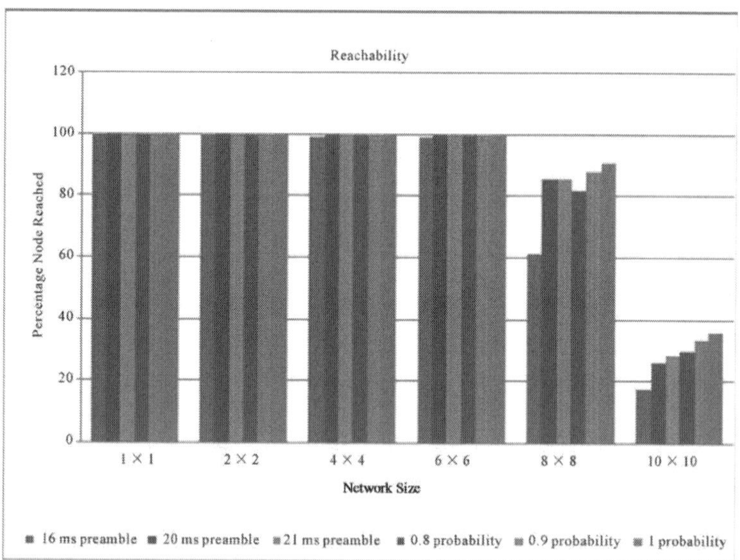

Figure 4. Percentage node reached vs network size.

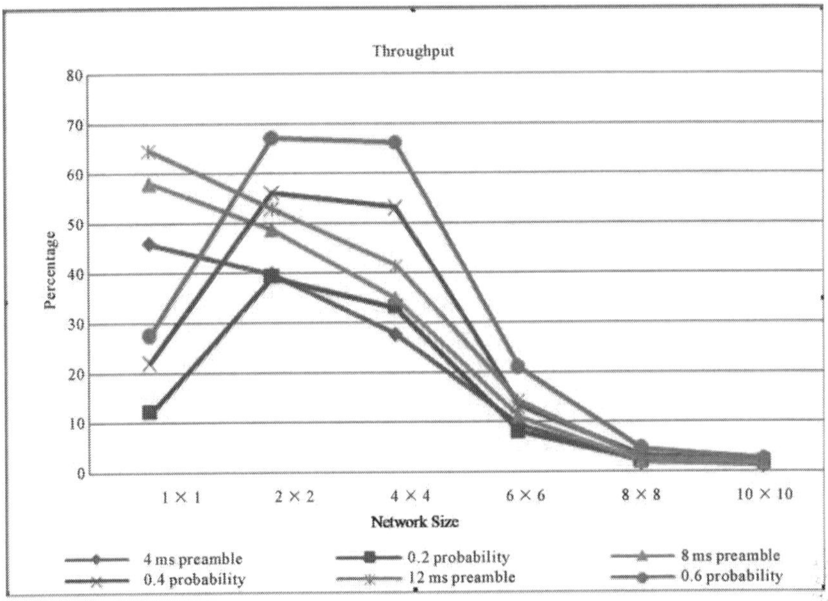

Figure 5. Percentage throughput vs network size.

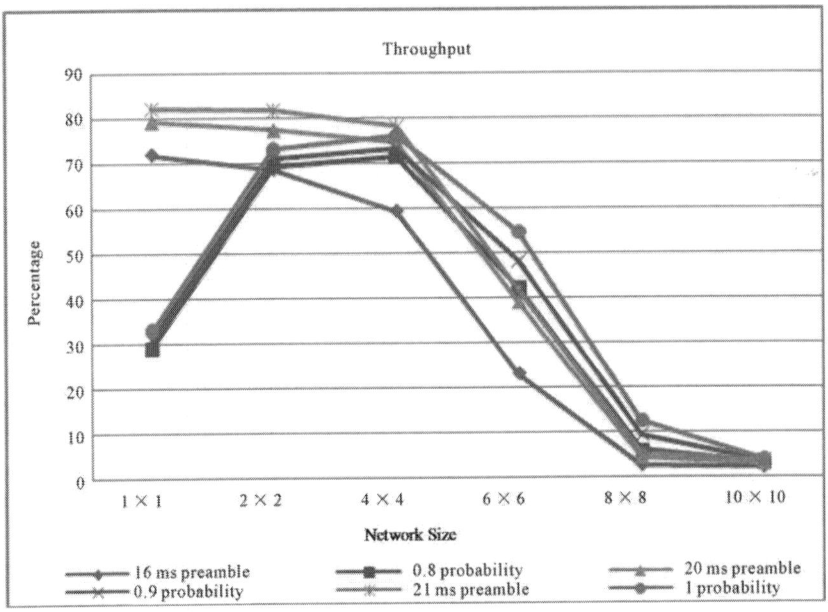

Figure 6. Percentage throughput vs network size.

5.3. Average Duplicate Packet

Figure 7 and Figure 8 show the duplicate packets/node versus network size, comparing the preamble-based broadcasting with the probability-based broadcasting. We observe that in a very dense area network (1 × 1) nodes received less duplicates packet when using probability-based broadcasting, due to the high number of collision and packet loss. We observe that in probabilitybased scheme, the duplicate packet reception is three to four times higher than the preamble-based scheme in various network sizes. As the network becomes sparse, nodes receives less duplicates packets in both the schemes, though in probabilistic scheme duplicate packet reception is higher.

5.4. Energy Consumption

Figures 9-11 show the energy consumption (mj/node) versus network size, comparing the preamble-based broadcasting with the probability-based broadcasting. We observe that energy consumption per node in probability-based scheme is always high in contrast to the preamble-based scheme. We conserve the energy per node in preamble-based broadcasting by allowing the node to go to dormant state when nothing is on the channel and save the energy at transmission node and receiver node by transmitting a small preamble and receiving a small preamble respectively before the data packet. It is observe that preamble-based scheme is best suited in term of energy conservation in the nodes as compared to the probability-based scheme. Energy consumptions per node are always high in probability-based scheme due to its always active node in the network and unnecessary energy consumption in nodes during the idle mode.

6. CONCLUSIONS

In this paper, we have proposed a preamble-based broadcasting technique for wireless ad hoc sensor networks. The proposed preamble-based broadcast technique is implemented and studied thoroughly, considering mobile nodes in wireless ad hoc sensor networks. An existing probability-based technique is also implemented and studied to compare with our preamble-based technique. We observe that our scheme provides high reachability in dense area networks and is a better solution in terms of reachability for dense area wireless ad hoc sensors networks. It is observed that our preamble-based scheme achieves significant throughput while consuming less energy and saving significant amount of duplicate packets.

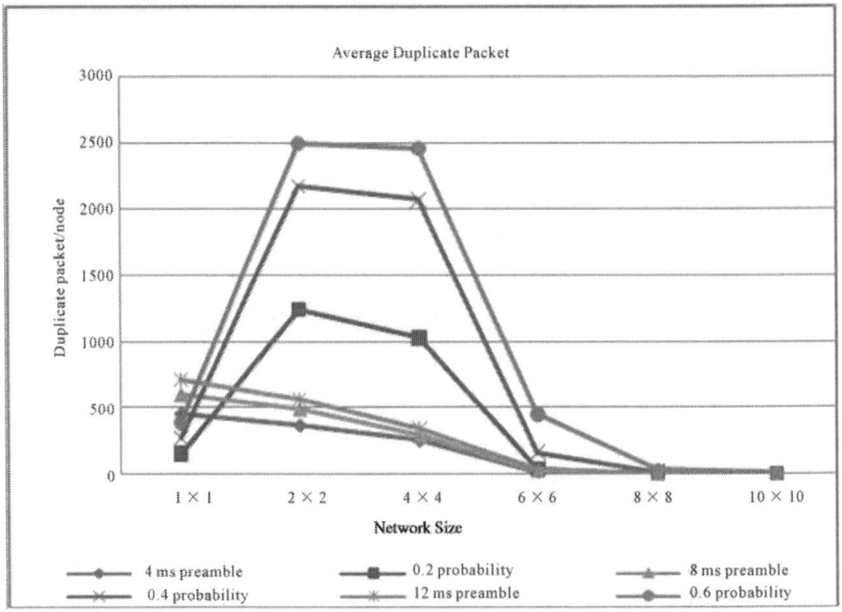

Figure 7. Duplicate packets/node vs network size.

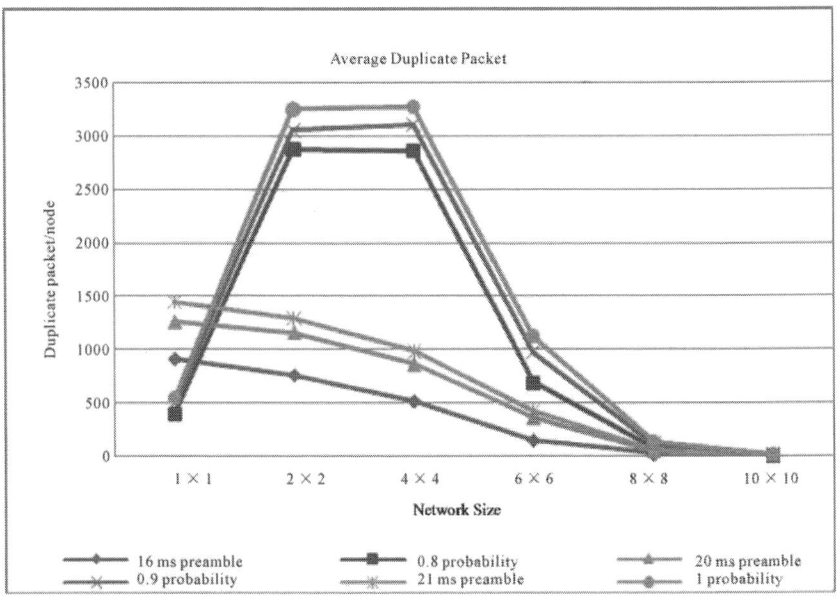

Figure 8. Duplicate packets/node vs network size.

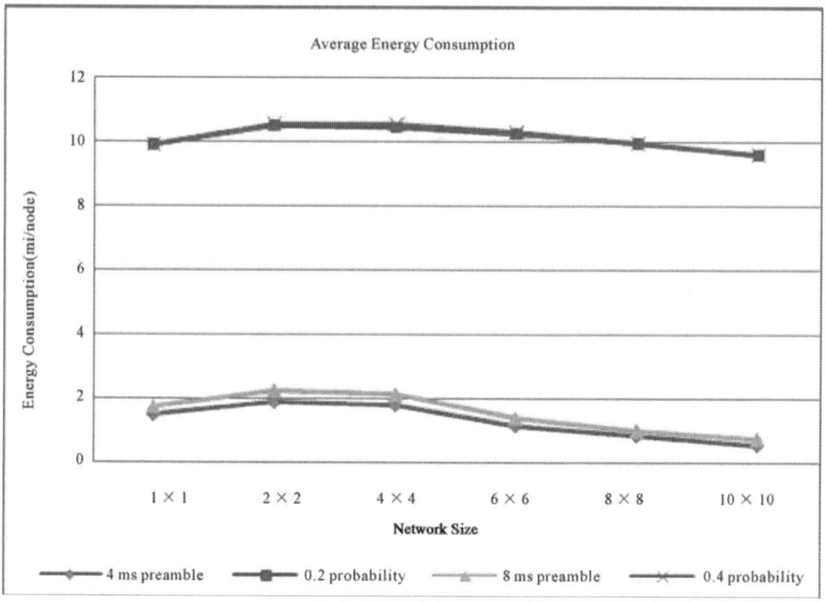

Figure 9. Energy consumption (mj/node) vs network size.

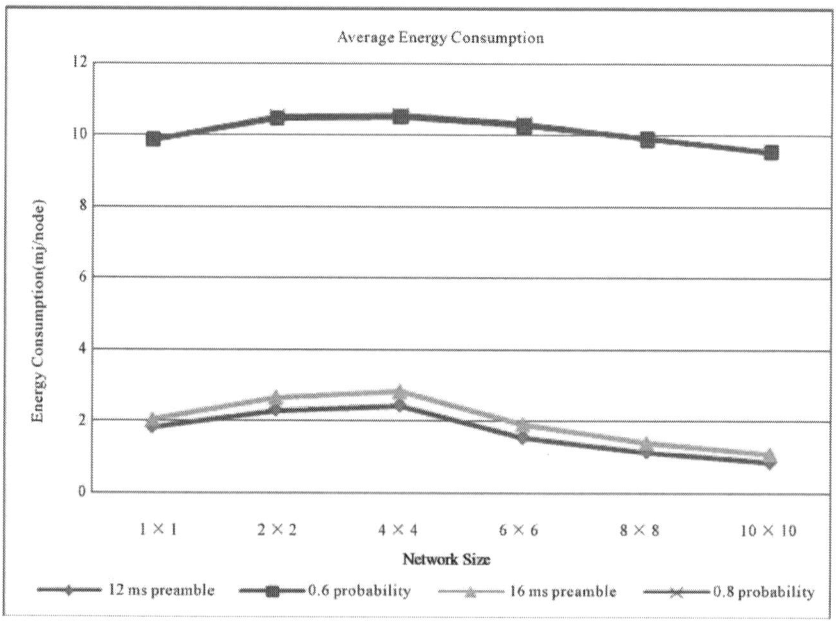

Figure 10. Energy consumption (mj/node) vs network size.

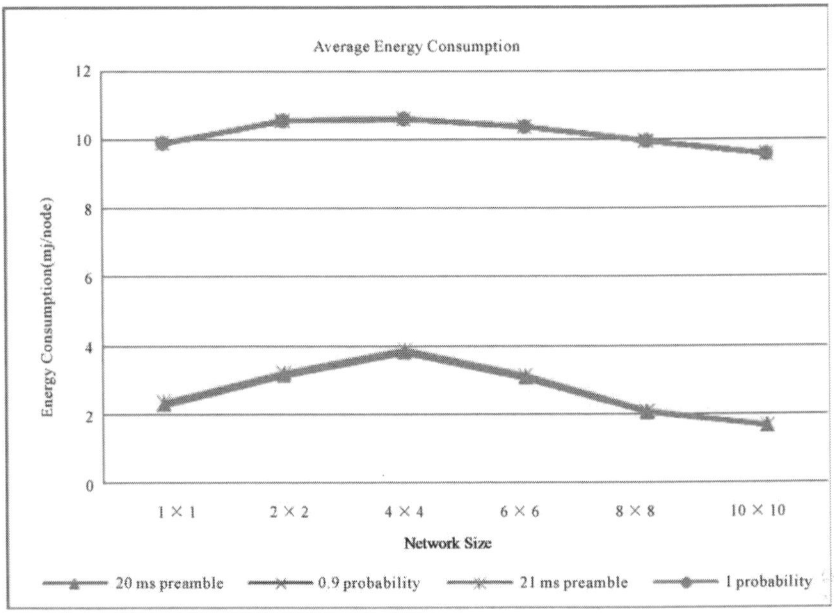

Figure 11. Energy consumption (mj/node) vs network size.

In future, we plan to make an analytical model for our proposed preamble based broadcasting technique in order to compare it with our simulation results. We also plan to make our preamble-based scheme dynamic, so that it can dynamically choose preamble size according to the extra area coverage and neighbor node information.

REFERENCES

1. T. He, S. Krishnamurthy, L. Luo, et al., "VigilNet: An Integrated Sensor Network System for Energy-Efficient Surveillance," ACM Transaction on Sensor Networks, Vol. 2, No. 1, 2006, pp. 1-38.

2. L. Selavo, A. Wood, Q. Cao, et al., "LUSTER: Wireless Sensor Network for Environmental Research," Proceedings of the 5th International Conference on Embedded Networked Sensor Systems, Sydney, 4-9 November 2007, pp. 103-116.

3. N. Xu, S. Rangwala, K. K. Chintalapudi, et al., "A Wireless Sensor Network for Structural Monitoring," Proceedings of the 2nd International Conference on Embedded Networked Sensor Systems, Baltimore, 3-5 November 2004, pp. 13-24.

4. Y.-C. Tseng, S.-Y. Ni, Y.-S. Chen, et al., "The Broadcast Storm Problem in a Mobile Ad Hoc Network," Wireless Networks, Vol. 8, No. 2-3, 2002, pp. 153-167.

5. B. Williams and T. Camp, "Comparison of Broadcasting Techniques for Mobile Ad Hoc Networks," Proceedings of the 3rd ACM International Symposium on Mobile Ad Hoc Networking & Amp; Computing, Lausanne, 2002, pp. 194-205.

6. W. Feng and L. Jiangchuan, "Duty-Cycle-Aware Broadcast in Wireless Sensor Networks," Proceedings INFOCOM 2009, 19-25 April 2009, pp. 468-476.

7. S. Guo, Y. Gu, B. Jiang, et al., "Opportunistic Flooding in Low-Duty-Cycle Wireless Sensor Networks with Unreliable Links," Proceedings of the 15th Annual International Conference on Mobile Computing and Networking, Beijing, 20-25 September 2009, pp. 133-144.

8. Y. Sun, O. Gurewitz, S. Du, et al., "ADB: An Efficient Multihop Broadcast Protocol Based on Asynchronous Duty-Cycling in Wireless Sensor Networks," Proceedings of the 7th ACM Conference on Embedded Networked Sensor Systems, Berkeley, 4-6 November 2009, pp. 43-56.

9. S. Yanjun, G. Omer and B. J. David, "RI-MAC: A Receiver-Initiated Asynchronous Duty Cycle MAC Protocol for Dynamic Traffic Loads in Wireless Sensor Networks," Proceedings of the 6th ACM Conference on Embedded Network Sensor Systems, Raleigh, 4-7 November 2008.

10. Kamra, V. Misra and D. Rubenstein, "CountTorrent: Ubiquitous Access to Query Aggregates in Dynamic and Mobile Sensor Networks," Proceedings of the 5th International Conference on Embedded Networked Sensor Systems, Sydney, 4-9 November 2007, pp. 43-57.

11. S. Nath, P. B. Gibbons, S. Seshan, et al., "Synopsis Diffusion for Robust Aggregation in Sensor Networks," ACM Transactions on Sensor Networks, Vol. 4, No. 2, pp. 1-40, 2008.

12. T. Instrument, "2.4 GHz IEEE 802.15.4/ZigBee-ready RF Transceiver," 2007. http://focus.ti.com/lit/ds/symlink/cc2420.pdf

CHAPTER 5

Applications and Security of Next-Generation, User-Centric Wireless Systems

Jerry Rick Ramstetter [1], Yaling Yang [2] and Danfeng Yao [3,*]

[1] Department of Computer Science, Rutgers University, Piscataway, NJ, 08854, USA

[2] Department of Electrical and Computer Engineering, Virginia Tech, Blacksburg, VA, 24061, USA

[3] Department of Computer Science, Virginia Tech, Blacksburg, VA, 24061, USA

ABSTRACT

Pervasive wireless systems have significantly improved end-users' quality of life. As manufacturing costs decrease, communications bandwidth increases, and contextual information is made more readily available, the role of next generation wireless systems in facilitating users' daily activities will grow. Unique security and privacy issues exist in these wireless, context-aware, often decentralized systems. For example, the pervasive nature of such systems allows adversaries to launch stealthy attacks against them. In this review paper, we survey several emergent personal wireless systems and their applications. These systems include mobile social networks, active implantable medical devices, and consumer products. We explore each system's usage of contextual information and provide insight into its security vulnerabilities. Where possible, we describe existing solutions for defending against these vulnerabilities. Finally, we point out promising future research directions for improving these systems' robustness and security.

KEYWORDS

wireless; pervasive; security; location based; context-aware; implantable medical devices

1. INTRODUCTION

The number of wireless devices for personal use (including mobile phones, wireless keyboards, and wireless networks) has increased dramatically in recent years. This change is in part due to decreased manufacturing costs and increased ease of deployment. Researchers are working on applying wireless technologies to emergent, user-centric applications (e.g. mobile social networking). These researchers believe that continued widespread adoption of wireless technologies will further improve end-users' quality of life.

We categorize wireless systems into three classes based on their communication and organizational infrastructure (or lack thereof). These categories are centralized, decentralized, and hybrid systems. In centralized wireless systems (e.g. wireless LANs), a base station or system administrator is responsible for managing the participating devices, including assigning security credentials and system resources to them. On the other hand, decentralized systems (e.g. peer-to-peer devices) lack pre-existing infrastructure and must organize themselves spontaneously.

We further categorize wireless systems as either contextual or traditional (e.g. non-contextual). Context-aware systems monitor their environment for changes and adapt to these changes. Context-aware systems can provide a more robust end-user experience, for example, by enabling the delivery of contextually relevant media and advertisements to end-users.

In this paper, we focus on the security and contextual aspects of pervasive, personal, often decentralized wireless systems. Decentralization creates technical challenges in the security and privacy of wireless systems due to a lack of trusted authorities therein. The personal nature of such systems dictates that successful attacks upon them may reveal highly sensitive information; this is especially the case for context-aware systems. Finally, the pervasive aspect of these systems gives attackers an ample number of devices to strike upon and to hide behind. As the adoption of user-centric wireless systems continues to grow, security vulnerabilities will create increasingly serious consequences for individuals. Part of the purpose of our review is to illustrate the importance of ensuring security in all user-centric wireless systems.

Our paper is organized as follows. We look at location or proximity based, user-centric wireless applications. In particular, we look at mobile social networks (section 2). We survey other types of user-centric devices, including consumer devices and implantable medical devices(section 3). We then provide a brief overview of cryptographic techniques, especially as it applies to wireless communications (section 4). As a general direction for improving location based services, we provide some discussion on the problem of location verification (section 5). In our conclusion, we discuss common trends and point out overarching directions for future work (section 6).

2. MOBILE SOCIAL NETWORKING

To begin our discussion of location-based services (LBS) for personal use, we turn to mobile social networking and the impact that contextual awareness might have on it. The phrase mobile social networking (MSN) is a broad category inclusive of all mobilely accessible social networking services; we include both context-aware and traditional social networking services here. In general, context-aware services are those that adapt according to environmental changes, including variations in physical location of use and changes in the end-user's intended application of use [1]. We consider traditional services to be services which do not fit the definition or intent of contextual awareness. Examples of traditional MSN services are Facebook, Myspace, and Twitter [2,3,4]. Though these services offer mobile access (e.g. via a WAP gateway and modern cellular phone), and can correlate end-user access with time, such services are incapable of correlating user access with intended use or physical location. System knowledge of detailed context information enables a more robust end-user experience. For example, services can recognize an end-user's repetitive use patterns, using those patterns to improve that user's socialization with other users. Unfortunately, highly available context information also allows for complex and intrusive attacks (such as violating a user's location privacy). We focus on context-aware mobile social networking (CAMSN) services in the following section.

Although service designers can apply context information to mobile social networking services in many ways, of the CAMSN services we have encountered, the primary goal has been to allow digital interaction between users concerned with the same physical location. For example, users might be able to create or edit content related to their shared physical location (e.g. reviews of the restaurant they are dining in). As another example, attendees at a social event might receive automatic notifications about persons in their vicinity with similar interests (the goal being to foster interaction between strangers).

2.1. Service Designs for Context-Aware Mobile Social Networking

We highlight some of the choices that context-aware mobile social networking service's designers make below.

Network architecture type: We find three architectural categories for CAMSN services: client-server, peer-to-peer, and hybrid. Client-server CAMSN services are centralized: their end-user platforms provide content by utilizing communications with fixed and authoritative application servers. Peer-to-peer CAMSN services are decentralized: they seek to avoid the usage of fixed servers or other authoritative entities. Peer-to-peer CAMSN services therefore need to spontaneously provide (1) localization or proximity functionalities, (2) network setup, and (3) application level functionalities. Hybrid architecture CAMSN services are a mixture of client-server and peer-to-peer architectures. In subsection 2.2, we describe in more details the network architecture types used by CAMSN services, and also provide examples of services using each.

Source of context and user information: A majority of CAMSN services derive application layer contextuality from either location information or proximity information. Location information refers to the knowledge of physical locations for all users (e.g. GPS coordinates) of a service, allowing the service to correlate and match user locations. On the other hand, proximity information for a user refers to knowledge of nearby users as determined via short range communications (e.g. Bluetooth). In subsection 2.3, we describe in more detail the methods with which CAMSN services can derive location or proximity information, and provide examples of services using each.

We find a further distinction among CAMSN services regarding where their user profile information is taken from. Most services require end-users to create a new, specific profile for usage on the service; these are non-augmenting services. Augmenting CAMSN services, however, allow users to augment existing, traditional social networking profiles (e.g. Facebook) with externally provided contextual information. An example of an augmenting CAMSN is WhozThat [5]. The data flow in WhozThat is bidirectional – information flows from end-user platforms to the WhozThat service (the IDs of proximate peers) and likewise flows in the opposite direction (the profile data of proximate peers). An augmenting CAMSN need not have this bidirectional data flow; CenceMe is one such example [6]. Client devices in this service push context information to their appropriate traditional social networking profiles. However, these same client devices do not receive context information from the service, they cannot answer the question "what devices are near me?".

2.2. Network Architectures for CAMSN Services

In this section, we explain in further details the network architectures used in context-aware mobile social networking services.

In client-server based CAMSN architecture, a client device might ask its application servers "what devices are near location X?" This same device may announce its location to its application servers. Most CAMSN services to date have client-server architectural design. These include FourSquare, BrightKite, Loopt, Google Latitude, and CenceMe [6,7,8,9,10].

CAMSN services in peer-to-peer architecture poll or discover physically neighboring devices without discovering location information for those devices. MobiSN, which defines an ad-hoc routing protocol and profile matching functionality, is one example of a completely peer-to-peer CAMSN service [11].

Hybrid architecture CAMSN services make use of some combination of client-server and peer-to-peer communications. A device on a hybrid service might discover proximity information (e.g. neighboring devices) via short-range communications, but use that information to look up user profiles on an authoritative application server. An advantage here is that the application server need not be concerned with location information. That is, an end-user device x will not ask the application server "what devices are near my location, which is L?" Rather, x would ask "what profile data is available for device y," where y is another device in the proximity of x (discovered via short-range communications). Examples of hybrid CAMSN services include WhozThat, Serendipity, and Aka-aki [5,12,13], all of which utilize both client-server (e.g. for user profile data retrieval) and peer-to-peer (e.g. for discovering proximity relations) network architectures.

2.3. Methods for Deriving Location or Proximity Information

We highlight the primary methods that context-aware mobile social networking services use to derive location or proximity information. Proximity information for a device refers to the knowledge of neighboring devices as locally discovered through short range communications.

Self reporting: CAMSN services of the self reporting type allow end-users to report their physical location to the service, for example by typing it in on their cell phone's keypad. This often cumbersome method of deriving location information relies on users to accurately (e.g. non-maliciously) report their location. We find that this method of deriving location information is rarely used, possibly due to the prevalence of GPS hardware on end-user platforms (e.g. cell phones). BrightKite is notable in that it allows users lacking a GPS enabled cell phone to self report their location [8].

Self-derived location: CAMSN services in this category use trustworthy, externally provided information to localize themselves. If the hardware platform used by such a CAMSN service is a modern cellular phone, the service might use GPS or Cellular Assisted GPS (A-GPS) to determine its location in real time (or at specified time intervals). Loopt and Google Latitude, for example, update A-GPS derived location information in real time, automatically dispersing that information to chosen friends [9,10]. FourSquare uses GPS derived location to suggest locations which users might check in to, thereby allowing them access to that location's content [7]. Where GPS coverage is not available, most of the A-GPS based services fall back to using cell tower triangulation to provide location information. We find that self derived location was not used in peer-to-peer CAMSN services, as gossiping GPS location data between devices associated with such a service would include neighbor discovery – effectively polling (see below).

Network-derived location: For CAMSN services in this category, the infrastructure or network is responsible for calculating a device's location and reporting this location back to the device. We did not encounter any CAMSN services that explicitly utilize network derived localization. This observation is likely because most CAMSN services to date have utilized cellular phone platforms, and such platforms generally include GPS receiver hardware. On the other hand, such platforms often use cellular-assisted GPS (A-GPS). Some implementations of A-GPS involve the wireless cellular network pushing additional information to receivers (e.g. cell phones) to aid in the localization process.

Peer polling and broadcast: CAMSN services utilizing polling continually look for neighbors over short range communication channels while simultaneously announcing their own presence over those channels. This method does not find location information for those peers but rather identifies which peers are nearby. Serendipity and WhozThat both utilize polling for neighbor discovery. Serendipity nodes broadcast and looks for Bluetooth IDs (BTIDs), whereas WhozThat (an augmenting CAMSN) utilizes user names from traditional social networks in place of BTIDs [5,12]. Other polling based CAMSN services are Jambo, which utilizes 802.11 WiFi rather than Bluetooth, and MobiSN [11,14]. Polling has the advantage of being inherently compatible with CAMSN services utilizing a peer-to-peer architecture, as polling does not require fixed infrastructure to work.

Hybrids of self reported location, self derived location, and proximity based mobile social networking services are possible. "Hybrid" as used here does not refer to a hybrid network architecture (see subsection 2.1), but rather to how a service derives location or proximity information. As mentioned, BrightKite is able to switch between self reporting and GPS derived location information. MobiLuck allows users to either check in to their locations or enable real-time GPS based updates, depending on their hardware platform and personal preferences [15]. Aka-aki uses a hybrid of peer proximity information (derived from Bluetooth) and A-GPS location information [16].

2.4. Attacks and Defenses in CAMSN services

In all types of mobile social networks, issues of security and privacy exist. Attacks on mobile social networking services (both contextual and traditional) include Sybil attacks, wormhole attacks, anonymity attacks, as well as social-engineering based phishing attacks. Furthermore, information leak and privacy issues are possible. Many of these problems can exist in a mobile social network without regard to the level of contextual awareness introduced by it. Below, we describe some of these problems in further detail and, where possible, provide mitigation strategies for them.

In a **Sybil attack**, a single entity takes on the roles of multiple entities (known as Sybil identities) for illicit purposes. The Sybil attack is possible in any service that utilizes reputation information assigned by service clients (e.g. peers of a device on a service), and is in many ways similar to the real-world concept of ballot stuffing [17,18,19].

Sybil attacks are possible in CAMSN. For example, an attacker can hijack a social networking site's highest ranked profiles list by creating many bogus accounts (or taking control of many legitimate accounts) and using those accounts to vote for a target profile. The bogus or stolen accounts need not cast their votes at the same time. As another example, an attacker using a CAMSN service can illicitly advertise a physical location (e.g. a bar or pub) by converging a number of bogus (or stolen) identities upon that location; the location will thereby appear popular to legitimate users of the CAMSN. In this case, the bogus or stolen accounts must converge on that physical location at the same time in order to make it appear popular.

Service designers and engineers can prevent a Sybil attack altogether only via careful certification of identities by a trusted authority [17], which must ensure every entity has only one identity. Zhang et al. designed a novel solution based on location-based cryptographic keys to prevent Sybil attacks, in particular, location-spoofing based Sybil attacks in wireless sensor networks [20]. The proposed location-based keys are generated using pairing-based identity-based cryptography by a trusted authority. Similarly, Trusted Computing Group (TCG) modules could allow for the attestation of code running on a MSN device [21], thereby helping to prevent the hijacking of legitimate identities. Power consumption of an on-board TCG module is an issue in the latter of these solutions, while the requirement of a trusted, external authority is unreasonable in the former (especially in the case of peer-to-peer systems). Methods of attesting memory contents in wireless sensor networks have been proposed [22,23]; these methods could potentially be applied to mobile social networks and the hardware thereof.

Steps a service's designers can take to limit a Sybil attack's effectiveness include limiting the privileges (or relative weight) of newly created identities and making the creation of new identities difficult. The latter can be accomplished, for example, by requiring that a CAPTCHA be solved [24]. Verifying the physical locations of identities to ensure that they move in a realistic manner has been proposed [19]. For example, the position of Sybil identities might be expected to shift locations rapidly. In the realm of CAMSN services, such location based methodologies for detecting Sybil attacks are of particular use when location information is already known – the cost of implementing them might be comparatively low (see this paper's discussion of Secure Location Verification in section 5). Lastly, a method of examining the connectivity characteristics of social graphs to prevent Sybil attacks has been proposed [25]. An overview of further solutions to the Sybil attack is found in [18].

Attacks against anonymity: CAMSN services are susceptible to attacks which leverage the fact that a user's identity is often intrinsically associated with a social networking profile and location information [26]. Such attacks attempt to violate the privacy of an end-user's location history by associating that history with a unique, application layer identifier. In peer-to-peer systems, such attacks are made possible by the exchange of unique identifiers (e.g. social networking service user names) and profile data between nodes. Malicious nodes can log these unique identifiers, allowing reconstruction of a target device's (or user's) location history.

In client-server systems, where unique identifiers are not swapped between nodes but rather from clients to application servers, direct anonymity attacks are possible by inserting malicious devices into the network. These malicious devices can continuously query an application server for the unique identifiers (or user names) of nearby devices and record the times at which such unique identifiers were seen. In both cases, attackers can use the observed identifiers to find corresponding social networking profiles. Importantly, anonymity attacks do not require a malicious node to receive information beyond what it would normally receive. That is, malicious entities can carry out such a direct anonymity attack by recording only a victim's unique identifier, that victim's physical locations, and the times at which the victim was seen (and correlating that information with the victim's profile).

Services can prevent or mitigate anonymity attacks via the use of encryption, though encryption is hard in the peer-to-peer case due the challenge of establishing shared keys (see section 4). A system using anonymous identifiers at the application level has been proposed [26]. This system is applicable to hybrid mobile social networking service architectures. It presumes that devices 1) utilize a short range communication technology for discovering proximity information (e.g. Bluetooth) and 2) have communication with a trusted identity server (e.g. via the Internet). In this system, no communications use unique identifiers. Rather, all communications use anonymous identifiers, which are assigned on a per-message basis by the trusted identity server. Only the identity server can map anonymous identifiers to specific devices, and the server will perform this mapping only after verifying that two potentially communicating devices are truly within each other's physical proximity. This system prevents malicious parties from mapping messages or profile information to a specific user.

Privacy leakage: While phishing attacks involve a malicious entity attempting to trick users, privacy leaks ostensibly occur with the user's permission. Both malicious and non-malicious entities can leverage this phenomenon. For example, a health insurance company might notice that a potential insuree's social networking profile is public, and look at that profile for information regarding the potential insuree's medical state. Though possible in any social network, the increased amount of personal information available to CAMSN services (and social networking services in general) dictates that such privacy leaks might have far greater consequences. For example, a website leveraging privacy leakage in CAMSN services has been released [27]. This website examines users' locations on CAMSN services and identifies users who are not in their home. By identifying users who are not at home, the site potentially aids malicious persons in finding empty homes to burglarize. The site only makes use of publicly available information; users who appear on the site have chosen to make their location publicly available (though, likely, without sufficient consideration). Sufficient documentation must be provided by mobile social networking services to ensure users are able to alter privacy settings. Default service privacy settings are of critical importance in battling privacy leakage, as even with documentation users might not understand how to alter privacy settings.

Network-based attacks: Mobile social networking services are susceptible to attacks on their infrastructure. Like many networked systems, mobile social networking services are subject to denial of service, eavesdropping, spoofing, and replay attacks. A denial of service attack consists of an attacker consuming excessive resources on a mobile social networking service's infrastructure. Twitter, for example, was brought down for days in August 2009 by a denial of service attack against its web servers [28]. An eavesdropping attack consists of a malicious user listening in on wireless communications, and is preventable via end-to-end encryption. A replay attack involves eavesdropping, recording, and replaying some critical pieces of information, such as a session authentication or other identifying message. Spoofing attacks occur when an attacker pretends to be an identity (or authority) which he is not. Replay attacks are often used in carrying out a spoofing attack. The link layer Address Resolution Protocol (ARP) is notoriously susceptible to spoofing attacks which allow an attacker to take control of the IP address of a legitimate network device [29].

The anonymous identifiers system proposed in [26] prevents eavesdropping, replay, and spoofing attacks. Services can use location verification (see section 5) to aid in the prevention of replay and spoofing attacks. Other methods of preventing replay attacks are summarized in [30].

A wormhole attack is a variant of the replay attack in which recorded data from a physical location A is replayed at a different physical location B, with the intent of making a user or device at location A appear to be at location B. Malicious entities can use wormhole attacks in mobile social networking services to create confusion amongst users or to alter network topology with the goal of maliciously intercepting (and replaying) traffic. Differing network speeds make wormhole attacks possible: in a peer-to-peer MSN, if device X sends a message m to device Y, m will follow an expected route (e.g. the fastest route) across peer devices. However, if a wormhole is introduced as the shortest path between X and Y, the message m might flow along the wormhole, thus allowing the wormhole's controller to eavesdrop on and manipulate the message m (assuming m is not encrypted). One solution to wormhole attacks involves the use of Packet Leashes, or limitations on a packet's lifetime and traveled physical distance [31]. A routing protocol which attempts to identify the effects of a persistent wormhole has been proposed [32]. Algorithms utilizing nodes dedicated to the task of monitoring network topology have been proposed [33]. Many other strategies to aid in the prevention and detection of wormhole attacks have been proposed and we refer the reader to the following literature for further reading [34,35].

3. CONSUMER PRODUCTS AND ACTIVE IMPLANTABLE MEDICAL DEVICES

Emergent user-centric and wireless systems consumer products may include personal media devices (e.g. an iPod paired with Bluetooth headphones), wireless computer mice, wireless keyboards, and off-the-shelf 802.11 WiFi hardware. As the cost of wireless transceivers continues to decrease, countless more devices and technologies will make use of them. The pervasive and often personal nature of these devices requires that designers pay attention to their security. We also discuss security issues associated with personal medical devices in this section.

3.1. Security and attacks in Consumer Products

We interpret the phrase consumer-product security in two ways. The first, which we adopt primarily, is that of protecting users from malicious or insecure devices. The second (and generally less common) view is that of protecting devices themselves from consumer attacks or "hacks" (for example, in order to access manufacturer disabled functionalities). A primary reason manufacturers intentionally disable functionalities is to enable product price tiering (e.g. low and high end products) while maintaining highly cohesive manufacturing processes: lower end devices will have excess functionalities disabled rather than removed altogether. These viewpoints of the phrase "consumer product security" are distinct but not mutually exclusive; many of the techniques proposed in dealing with the second viewpoint are directly applicable to the first. For example, the use of secure boot processes, robust code reviews, and component isolation in consumer devices has been advocated in order to prevent the introduction of unauthorized, feature-enabling new code [36]. These measures have direct end-user security benefits in that they aid in preventing the introduction of all new code, whether that code be user intended or malicious.

In the remainder of this section, we discuss security vulnerabilities in consumer products, describing instances of successful attacks upon those vulnerabilities where possible.

Consumer products are susceptible to attacks against anonymity, similar to CAMSN. One example of this is the Nike+iPod Sport Kit [37]. The product utilizes two paired devices, a sensor and a receiver. The sensor is placed in an end-user's shoes, while the receiver is attached to the user's pre-existing iPod. The users' footsteps are monitored by the sensor and reported to the receiver, allowing the user to monitor her exercise. The product utilizes a non-encrypted unique ID for every sensor. This allows attackers to sniff for the system's presence and, in turn, monitor the end-user's presence at various physical locations. Because the system is physically based (e.g. users wear it), attackers can easily associate overheard unique IDs with individual persons. In this case, the manufacturers of the device failed to recognize that communications between two statically paired devices (sensor and receiver) could easily be encrypted with a symmetric cryptography system.

Side channel attacks allow an attacker to gain useful information by means other than message contents; these attacks are a type of information leak. Examples of side channels are inter-packet timing, packet header contents, packet size, and network flow size. An attacker can observe all of these without knowledge of message contents or protocol encryption schemes. Side channel attacks are possible in any networked system including mobile social networks, vehicular ad-hoc networks, and medical devices. Certain features of consumer devices make them particularly favorable targets for side channel attacks. Specifically, side channel attacks are most easily carried out when an estimate of the type of network traffic to be eavesdropped is available.

The SlingBox Pro, a wireless consumer media device utilizing Variable Bit Rate (VBR) compression, was found to suffer from information leak problems [37]. The SlingBox Pro's goal is to encode live media signals (e.g. television or movies), apply VBR compression to the encoded media, and stream the compressed media over a network to the Internet. In testing the SlingBox Pro, researchers used an encrypted 802.11 WiFi network for communications. Researchers discovered that they did not need to examine message contents (nor break 802.11 WiFi security measures) in order to discover the movies being viewed by SlingBox Pro users. Rather, the SlingBox's use of VBR compression allowed investigators to correlate network flows with individual media items. For an observed network flow, researchers specifically correlated packet size, inter-packet timing, and packet rate with individual media items – all without knowledge of packet contents. We note that the highly specific purpose of the SlingBox Pro eased the task at hand for two reasons. First, researchers knew apriori the type of traffic they'd be concerned with (e.g. encoded and compressed movies). Secondly, the device's specific purpose ensured that researchers did not need to filter out extraneous traffic (e.g. web downloads) during their sniffing and analysis.

Obvious solutions for preventing side channels include padding and attenuating signals. Padding emitted signals with random noise would prevent that signal from being identified; however, issues of energy consumption remain. For example, the aforementioned attack against the SlingBox Pro could have been prevented by padding the VBR compressed media streams with random data. The problem here is that the goal of VBR compression is to reduce bandwidth usage, whereas padding media streams with random data would increase bandwidth usage. Signal attenuation would use metal shielding to prevent wireless signals from crossing physical boundaries. The economics of massive scale attenuation are hard to justify, as it might require wrapping whole buildings in shielding.

Device fingerprinting attacks occur when side channels are used to fingerprint (or uniquely identify) a wireless device. Such unique identification can be used to track a device even if the device takes proactive steps to ensure its anonymity. By simple measurement of a device's physical layer signal characteristics (such as the signal strength, angle of arrival, and time of arrival), invasive localization techniques might be able to track a wireless device's location without appropriate end-user consent, thereby violating the user's location privacy. Because such device fingerprinting is independent of MAC and IP addresses, malicious user tracking may be possible even if a device takes proactive steps (such as frequently changing its MAC and IP addresses) to ensure its anonymity. Fingerprinting attacks can be mitigated by introduction of random communication delays, use of periodic transmissions, introduction of signal attenuation, and use of randomly varying resistors in transceiver hardware (e.g. to randomly alter that signal's physical layer characteristics) [38,39].

Other common methods for protecting location privacy in the face of such device fingerprinting attacks are based on the concepts of mixed zones and anonymous identifiers [40,41,42]. In these methods, neighboring wireless devices sometimes enter a completely silent state wherein nothing is transmitted by any device. This silent state can be entered whenever a specified time period elapses or whenever a device enters a certain geographic area. At the conclusion of the silent state, all wireless devices adopt new identities (e.g. new MAC addresses and IP addresses). Since during the silent period the localization system cannot track mobile devices, it may be hard for a localization system to associate the new device identities and locations with the old (e.g. before the silent period) device identities and locations. In this way, the silent state creates a mixed zone that breaks any attempt by a localization system to continuously track a user's location, and hence protects the location privacy of users. Unfortunately, such methods have severe implications in that they limit the ability of all localization systems to function– even trusted ones. Similarly, the long disruptions on communication introduced by such methods may be unacceptable.

Many further attacks are possible on consumer products. These attacks include spoofing, phishing, eavesdropping, and wormholes, all of which are described in detail in subsection 2.4. Some manufacturers have proposed strategies for low cost improvements to consumer device security. For example, the usage of a low cost IPSEC proxy to encrypt communications from consumer devices has been proposed [43].

3.2. Security Trade-offs in Active Implantable Medical Device Designs

In this section, we focus on active implantable medical devices (AIMDs), a category inclusive of pacemakers, implantable cardiac defibrillators (ICDs), and implanted drug delivery systems. Such devices are becoming increasingly technologically advanced; current generation devices already support remote monitoring by health professionals via 802.11 WiFi [44]. Although external personal medical devices (e.g. glucose meters) and passive implantable medical devices (e.g. replacement joints) are likewise advancing technologically, we do not consider them. Because AIMDs can be repaired or replaced only via highly invasive surgical procedures, and because technologically advanced AIMDs might be susceptible to network attacks, interesting choices must be made and often competing goals must be balanced in designing them.

Designers of AIMDs must ensure that device security and patient privacy are maintained during attacks. However, design techniques aimed at addressing the goals of security and privacy are often at odds with other design goals. We highlight some of these trade-offs below, and, where possible, highlight some promising research in addressing them. In the following paragraphs, we use the term security to refer to the information security principles of confidentiality, integrity, and availability. We do not concentrate on physical safety, except in the context of AIMDs carrying out their primary, possibly life preserving functionalities.

Security vs. energy usage: Designers must keep the power usage of AIMDs to a minimum, especially considering the physical inaccessibility of AIMDs [45]. This places constraints on the computational power and communications abilities of AIMDs. For example, asymmetric cryptography may not be suitable for use in AIMDs due to its computational requirements.

Current-generation AIMD bearers commonly carry bracelets or other forms of identification to alert medical professionals of the AIMD's existence (e.g. in an emergency situation). Making these identifying bracelets or cards computationally powerful, and pairing them on a one-to-one basis with their AIMD, has been proposed [46]. An AIMD can offload long range communication and intensive processing onto an externally paired device. The one-to-one pairing between an AIMD and its external device allows the use of power-saving symmetric cryptography between the two. The lack of severe energy constraints on the external device allows for the use of computationally more-expensive asymmetric cryptography between it and other external devices (e.g. programmers).

Certainly other power-saving measures are possible. Medical practitioners and AIMD designers must take care to ensure that communications with an AIMD have sufficient benefit to warrant energy usage. For example, medical practitioners might allow non-critical telemetry data to queue on an AIMD rather than fetching it in near-real time, thus avoiding unnecessary AIMD energy usage.

Security vs. device lifetime: AIMDs cannot be replaced or physically repaired without surgery. Similarly, they cannot be updated (e.g. software patching) without a patient visit to a trusted device programmer (e.g. physician). Because of these facts, AIMDs must be designed with longevity in mind. This requirement is in contrast to the security paradigm of, for example, desktop computers, which require frequent rollouts of software updates to correct security flaws.

Protocols used to communicate with AIMDs must be designed for a very long lifespan. A single AIMD might serve a patient for 15 or more years, and this single AIMD might communicate with many different types and generations of external hardware throughout its lifespan. Each of these pieces of external hardware must support communication with the possibly obsolete AIMD; communication with a functioning AIMD cannot be hindered in the name of technological obsolescence [47].

Security vs. accessibility: An AIMD must be secure against an attacker on a day-to-day basis, but in an emergency situation that same device must allow for accessibility by previously unauthorized personnel (e.g. emergency room physicians or paramedics) [45]. An AIMD cannot easily differentiate between these scenarios in order to vary security settings, this due to a lack of physical access to the AIMD and aforementioned power constraints on it. Designers of an AIMD, then, might consider criticality-aware access policies infeasible (see [48] for reading on criticality-aware access policies).

Solutions for secure AIMD accessibility which are completely internal to an AIMD (that is, without a paired external device) have been proposed. One such solution proposes that AIMDs emit an audible warning during communication sessions, thereby enabling a patient to physically leave the communication range of unauthorized communicators [46]. It has been proposed that AIMDs and their programmers can mutually authenticate one another via observed physiological signals (for example, heart rate variability). These mutually observed signals would be used as the basis for a biometric encryption and authentication scheme between programmer and AIMD [49,50].

Security vs. functionality: As computational and medical hardware continue making strides in power efficiency, it is possible to put more functionality on-board an AIMD. Designers should analyze these functionalities to ensure they do not introduce excessive security or patient safety risks. For example, the ability to reprogram or otherwise communicate with an AIMD from a long distance is of great benefit medically, but unfortunately also allows attackers to carry out malicious activities from a long distance. Similarly, strong authentication between AIMDs and communicating devices is desired from a security standpoint, but latencies introduced by such a system may be medically unacceptable [46].

Security vs. data retention: Though a medical practitioner might desire an AIMD to contain a patient's entire medical history, this would allow a successful attacker access to the patient's history. Designers should evaluate the trade-offs between inclusion of data, the desire to keep that data out of an attacker's hands, and the effectiveness of device security measures.

Beyond the trade-offs of AIMD design described above, specific attacks on AIMDs are possible, including attacks against anonymity. Any communication system deployed in an AIMD must guard against denial of service attacks. Such attacks might include blocking of computational resources, depletion of energy reserves, or jamming of communications channels [45]. An AIMD utilizing a challenge-response authentication protocol might be susceptible to all three of these defects: generating a response utilizes computational resources, wireless communication utilizes significant energy reserves, and that same wireless communication further consumes communications channel bandwidth. An AIMD can mitigate these attacks by filtering communication from devices which are known, or reasonably suspected, to be malicious [51,52]. Efficiently and accurately categorizing third party devices as malicious remains a challenge.

4. Cryptography and Wireless Security

Cryptographic techniques form the first line of defense for securing wireless communications. Although this paper is not particularly focused on applied cryptography in wireless systems, we give a brief overview of the basic concepts. The three basic security requirements specified by the IEEE for wireless local area network (WLAN) environments are authentication, confidentiality, and integrity. Authenticity is to ensure the identity of the sender of a message. Confidentiality guarantees that a message is kept secret and can only be read by the intended receiver. Finally, integrity ensures that any tampering or distortion of a message by adversaries is identifiable. Many cryptographic protocols have been proposed to secure communications in wireless networks.

There are two main cryptographic paradigms: symmetric and asymmetric. Symmetric cryptographic systems (such as the Advanced Encryption Standard, AES [53]) use the same secret key for message encryption and decryption. They require the sender and receiver to agree on a shared key before communicating. Symmetric encryption is fast, relatively simple, and widely used for encrypting large data flows. Asymmetric encryption schemes (such as the RSA public-key encryption scheme [54]) require the receiver to have a key pair—a public key known by others and a private key known only to the receiver. The sender encrypts a message with the public key of the receiver. The receiver, upon obtaining the ciphertext, decrypts it with her private key. In public-key signature scheme, a signer uses her private key to sign a message for message integrity purpose and produces a digital signature. A verifier, upon obtaining the signature and the message, uses the signer's public key to verify the digital signature and make sure that the message has not been tampered with.

Asymmetric or public-key encryption schemes require the sender to know the correct public key of the receiver, similarly public-key signature scheme requires the verifier to know the public key of the signer. The authenticity of a public key is extremely important—using the wrong public key in decryption may result in the receiver erroneously accepting an adversary's message, possibly leading to attacks. Asymmetric cryptography is computationally more expensive than symmetric ones, making its usage in embedded or power constrained environments difficult.

In an environment where there are central trust authorities, services can use digital certificates to bind a public key to an identity and verify that a public key belongs to an individual. For example, VeriSign is one of the Internet's major certificate authorities. The digital certificates used in SSL and HTTPS utilize public key cryptographic schemes. Specifically, they use a public key digital signature scheme for initial trust establishment. Once trust is established between the client and the server (e.g. a Web server and a client's browser in the HTTPS case), a shared key is negotiated and used for encrypting subsequent communications.

In decentralized wireless networks (such as wireless ad-hoc networks) the infrastructure is fluidic—there is no pre-defined trust authority and the rate at which nodes join and leave the network may be high. The task of establishing trust between nodes is therefore difficult. For example, in Mobile Ad-Hoc Networks (MANETs) each node voluntarily forwards data to other nodes, but the determination of which nodes forward what data is made dynamically based on network connectivity. The mobile nature of MANETs leads to frequent network topology changes. Securing the communication in such a decentralized wireless system is challenging and may require advanced cryptographic tools such as threshold cryptography [55,56]. In threshold cryptosystems, a secret (e.g. certificate or other valuable data) is split into shares, and any k nodes can combine their shares to recover the master secret, which may be used for signing certificates or other critical operations. The advantage of using threshold cryptosystems in MANET is that no single node needs to take the responsibility of storing the master secret and being available and reachable all the time. The distribution of secrets among a number of nodes effectively decentralized the trust management without compromising the security. We refer readers to the literature for more discussion on applying threshold cryptography to securing decentralized wireless systems [57,58,59,60].

5. INCREASING ROBUSTNESS WITH LOCATION VERIFICATION AND COMPUTATION

The use of localization services in pervasive devices is increasing. Many technologies have been built upon an underlying assumption of available location information. These applications may have a need to securely verify the locations reported by pervasive devices, as such pervasive devices may be malicious or simply incorrect in reporting their location. Inaccuracies, whether malicious or accidental, are possible without regard to who or what is performing the localization. When a device is responsible for calculating its own location using trusted, externally provided information (as in self-derived localization), the device may calculate and report a malicious or incorrect location. Likewise, when a device has its location calculated by trusted infrastructure, and this location is then reported back to the device (as in network-derived localization), the device might maliciously alter information used by the network in calculating its location. For example, a malicious device can alter its signal characteristics.

The goal of location verification is to verify that location information reported by potentially malicious devices is accurate to some degree; service designers should view location verification as an additional, security-enhancing step in the process of localization. Location verification is a distinct problem from secure localization; the latter's goal is to allow devices to localize themselves in a hostile environment [61]. We do not consider the problem of secure localization in this paper.

5.1. Distance Ranging

Distance ranging is used to enable location verification via distance bounding, i.e., distance ranging is the foundation of the location verification. In a distance bounding location verification system, one device (a verifier) asserts that another device (a prover) is within a preset distance from it. Below, we provide some insight into the methodologies used to carry out distance ranging.

Time of Flight (TOF) ranging methods entail precise measurement of the travel time t of data between two devices, A and B [62]. A ranging system can calculate the maximum distance from A to B by knowing the communication medium's transmission speed and an estimate h of hardware induced delays. In the case of radio transmissions, where the transmission speed is c (the speed of light), the maximum distance would be $c(t-h)/2$.

Time Difference of Arrival (TDOA) ranging systems use multiple TOF verifiers to triangulate the location of a verifier in a process known as multilateration [63,64]. Ranging via TOF does not require bidirectional communication; a unidirectional data transmission is sufficient for the receiver (e.g. verifier) to calculate its distance from the sender (e.g. prover) provided packets are appended with their send time and clocks are synchronized.

Malicious parties can abuse TOF ranging in both the unidirectional and bidirectional cases. In the unidirectional message case, the receiver cannot guarantee that the send time appended to the received packet is accurate. In the bidirectional message case (a message from A to B, then from B to A), A cannot guarantee that B did not delay its return message unnecessarily. As B can only increase the range estimates by delaying the return message, verification systems can still function [65,66]. Even operating under the assumption that both parties are trusted, TOF ranging requires highly precise timing measurements, and is thus subject to large discretization error [67]. A clock accurate to millisecond precision will allow for up to 300 kilometers of error in distance calculated during a round trip transmission (assuming radio transmissions are used). It is possible, though costly, to use alternative communication mediums with a slower propagation speed in order to allow for lesser clock accuracy. The use of ultrasound in one direction of communication lowers the error term by about the order of 106. Major problems with the usage of such slower mediums include their high cost and their susceptibility to wormhole attacks.

Received Signal Strength (RSS) ranging utilizes the fact that radio or ultrasound signals decay with distance; receipt of a weak signal indicates either a weak transmitter or a large distance from the sender. Knowing the transmitter's output power in advance allows for ranging by mapping the received signal power through a function to distance. Many technologies exist to utilize existing 802.11 WiFi or Bluetooth infrastructure to carry out RSS ranging. Unfortunately, a malicious node can defeat RSS ranging by varying its transmit power in different directions (e.g. by using directional antennas), or by changing its transmit power with time. In addition, off-the-shelf, commodity hardware often does not make available an accurate measure of received signals' power. 802.11 WiFi device drivers are not required to provide access to this information through device drivers. Rather, the drivers must specify a RSSI value, which need not linearly map to received signal power in decibels [68].

5.2. Methods for Location Verification

One of the earliest solutions to the problem of secure location verification is distance bounding, which attempts to prove that a pervasive device, a prover, is within a bounded distance of a trusted verifier [65,66]. The physical space in which a verifier will attest to a prover's location is known as that verifier's Region of Acceptance, or ROA. The distance bounding that a verifier performs on a prover involves distance ranging. The ability of a verifier to perform its job is based on the assumption that either 1) the prover has only limited capability to distort the distance ranging process (e.g. the prover can increase but not decrease his calculated range from a verifier) or 2) the prover does not know the location of verifiers.

Probabilistic methods for location verification have been proposed [69,70]. These methods utilize the fact that it is possible to probabilistically relate the hop count (or number of devices through which a message passes) between nodes of a randomly dispersed Wireless Sensor Network (WSN) to Euclidean distance [71]. Namely, they work by bounding the Euclidean distance between two nodes according to the number of network hops between them. Unfortunately, these methods impose sufficiency limitations on network density; malicious or malfunctioning nodes are more difficult to detect in sparse networks.

Attacks against such probabilistic methods of location verification are possible. As the methods rely on reputation information communicated between peers, an attacker can use a variant of the Sybil attack to disrepute a victim node and cause blacklisting of that node. Similarly, a malicious node can alter the per-packet hop count information on which probabilistic location verification methods are dependent. Methods of securing verification systems against these and other attacks have been proposed. Inferring hop count from packet length might help in securing networks against nodes which maliciously alter the hop count of packets [69].

Two algorithms for secure location verification have been proposed in [72]. These algorithms attempt to verify that estimations of the same parameters (e.g. node location) are sufficiently close. The first algorithm looks for inconsistencies in four derived matrices (representative of network layout information), while the second is an iterative algorithm which utilizes an indicator value based on node neighbor consistency.

Neighboring sensors may be utilized for location verification. A method of position verification tailored to Vehicular Ad Hoc networks (VANETs) has been proposed [73]. This method utilizes a weighting of multiple sensors. Importantly, in the work's context, sensors refer to parameters that are used to heuristically determine whether or not a node is reporting the wrong location. As an example of the work's defined sensors, the acceptance range sensor utilizes the fact that VANET transceiver hardware typically has a well defined range; communications received from nodes known to be outside this range are flagged as likely to be lying about their position. As another example, the proactive exchange of neighbor tables sensor communicates neighboring node information across nodes, using it to find and flag inconsistencies in the reported network topology. Another location-based neighborhood authentication scheme was proposed by Zhang et al. that takes advantage of a group of special mobile anchors [74]. The anchors may be mobile robots and can perform coordinated group movement for collecting and measuring sensor data for location verification.

5.3. Robust Location Computation for Attack Source Positioning

This section concerns methods for locating attack sources and attackers. With the pervasive deployment of wireless systems, it is quite easy for an attacker to launch network attacks from a wireless terminal against remote critical network infrastructure—including national, financial, energy, transportation and military systems. Traditional trace-back techniques for wired networks can only identify the wireless network access point used by an attacker. Such rough identification is far from sufficient in wireless networks, as the attack source can be any node in the coverage range of the access point. The highly mobile, anonymous, and stealthy nature of wireless communication demands that finding the true location of an attacker be done using wireless localization schemes.

Traditional localization methods [75,76] rely on passive observation of the attacker's signal characteristics, such as its connectivity with neighboring access points, signal strength, etc., to make a position estimation. An intelligent attacker equipped with advanced radio technologies, like directional antennas and software defined radios (SDRs), can change its beam (or directional radio) direction and radio parameters to distort these signal features as well as significantly limit the measurements of its signal. In this way, it becomes very difficult for a localization system to compute the accurate location of an attacker and hence such an attacker can hide his/her position with great ease and anonymity.

A method for enabling the robust location computation of an intelligent attack source is proposed in [77]. In this method, the network proactively coordinates the wireless access points around the attacker to lure the attacker to change its beamforming (or directional signal filtering) direction. In this way, the attacker unintentionally allows more access points to measure its signal features, such as angle of arrival, Received Signal Strength Indicator (RSSI), and connectivity to its neighboring nodes. With enough measurement of network topology and wireless transmission features, the attacker can be correctly positioned.

6. CONCLUSIONS AND OPEN PROBLEMS

We have described the challenges faced in securing the communication and services of user-centric wireless systems. Such systems are often decentralized. On one hand, this decentralization enables users to conveniently share, access, and control unprecedented amounts of information, resources, and services. On the other hand, the broadcast and ad-hoc nature of wireless communication creates security and privacy challenges. We started our discussion by looking at location or proximity based, user-centric wireless applications. In particular, we explored the security ramifications of decentralization and contextualization on mobile social networks. We discussed location verification techniques as a method of improving these and other location based services. We looked at security issues that must be dealt with when using wireless technologies in consumer products and active implantable medical devices.

At present, many methods for protecting location privacy against a malicious localization or tracking system depend on clients and networks co-operating in a security protocol. Network trust cannot always be established, however (e.g. at a public access point). It will be important to develop solutions that allow users to control their location privacy as well as their own communication history without establishing network trust. Furthermore, current methods for protecting location privacy might make the localization process difficult for all localizers—even trusted ones. Contextually supporting pervasive devices with location information while simultaneously preserving location privacy remains an issue.

Preventing information leaks via side channels remains an issue. Many wireless devices are susceptible to side channel attacks, for example by fingerprinting of a wireless transceiver's physical layer characteristics. Current methods of preventing against information leaks are either expensive or obtrusive. As wireless technologies become more pervasive, more opportunities for observing leaked information will be made available to attackers. It will be necessary to prevent information leakage at all communication layers, including the hardware level.

Many current cryptographic solutions (e.g. public key cryptosystems and threshold cryptography) consume a great deal of energy, and as such are unsuitable for resource constrained devices. Although a device's resource constraints can often be worked-around (e.g. via the use of a paired, non-resource-constrained device), security measures completely internal to that device would have significant benefits, including lower overall energy usage and fewer system components.

ACKNOWLEDGMENT

This work has been supported in part by NSF grant CNS-0831186.

REFERENCES

1. Schilit, B.; Adams, N.; Want, R. Context-Aware Computing Applications. In Proceedings of WMCSA '94: The 1994 First Workshop on Mobile Computing Systems and Applications, Santa Cruz, CA, USA, 8–9 December 1994; IEEE Computer Society: Washington, DC, USA, 1994; pp. 85–90.

2. Facebook. Available online: http://facebook.com (accessed on 22 January 2010).

3. MySpace. Available online: http://myspace.com (accessed on 19 January 2010).

4. Twitter. Available online: http://twitter.com (accessed on 19 January 2010).

5. Beach, A.; Gartrell, M.; Akkala, S.; Elston, J.; Kelley, J.; Nishimoto, K.; Ray, B.; Razgulin, S.; Sundaresan, K.; Surendar, B.; Terada, M.; Han, R. WhozThat? evolving an ecosystem for context-aware mobile social networks. IEEE Network 2008, 22, 50–55.

6. CenceMe. Available online: http://cenceme.org (accessed on 22 January 2010).

7. Foursquare: Learn More. Available online: http://foursquare. com/learn_more (accessed on 22 January 2010).

8. BrightKite: Learn More. Available online: http://brightkite. com/learn_more (accessed on 22 January 2010).

9. About Loopt. Available online: http://loopt.com/about (accessed on 22 January 2010).

10. Google Latitude. Available online: http://latitude.google.com (accessed on 22 January 2010).

11. Li, J.; Khan, S.U. MobiSN: Semantics-based Mobile ad hoc Social Network Framework. In Proceedings of 52nd IEEE Global Communications Conference (Globecom), Honolulu, HI, USA, December 2009.

12. Eagle, N.; Pentland, A. Social Serendipity: Mobilizing Social Software. IEEE Pervasive Comput. 2005, 4, 28–34.

13. Aka-Aki. Available online: http://aka-aki.com (accessed on 22 January 2010).

14. Jambo. Available online: http://jambo.net (accessed on 22 January 2010).

15. Mobiluck. Available online: http://mobiluck.com (accessed on 22 January 2010).

16. Aka-Aki Blog. Available online: http://blog.aka-aki.com/?p=237 (accessed on 22 January 2010).

17. Douceur, J.R. The Sybil Attack. In IPTPS '01: Revised Papers from the First International Workshop on Peer-to-Peer Systems; Springer-Verlag: London, UK, 2002; pp. 251–260.

18. Levine, B.N.; Shields, C.; Margolin, N.B. A Survey of Solutions to the Sybil Attack; Techical Report 2006-052; University of Massachusetts Amherst: Amherst, MA, USA, 2006.

19. Newsome, J.; Shi, E.; Song, D.; Perrig, A. The sybil attack in sensor networks: analysis & defenses. In Proceedings of IPSN '04: The 3rd

International Symposium on Information Processing in Sensor Networks, Berkeley, CA, USA, 26–27 April 2004; ACM: New York, NY, USA, 2004; pp. 259–268.

20. Zhang, Y.; Liu, W.; Lou, W.; Fang, Y. Location-based compromise-tolerant security mechanisms for wireless sensor networks. IEEE J. Sel. Area. Commun. 2006, 24, 247–260.

21. Trusted Computing Group. Available online: http:// trusted computinggroup.net (accessed on 22 January 2010).

22. AbuHmed, T.; Nyamaa, N.; Nyang, D. Software-Based Remote Code Attestation in Wireless Sensor Network. In Proceedings of 52nd IEEE Global Communications Conference (Globecom), Honolulu, HI, USA, December 2009.

23. Shaneck, M.; Mahadevan, K.; Kher, V.; Kim, Y. Remote software-based attestation for wireless sensors. In Proceedings of The 2nd European Workshop on Security and Privacy in ad hoc and Sensor Networks, Visegrad, Hungary, 13–14, July 2005.

24. von Ahn, L.; Maurer, B.; McMillen, C.; Abraham, D.; Blum, M. reCAPTCHA: Human-Based Character Recognition via Web Security Measures. Science 2008, 321, 1465–1468.

25. Yu, H.; Kaminsky, M.; Gibbons, P.B.; Flaxman, A. SybilGuard: defending against sybil attacks via social networks. SIGCOMM Comput. Commun. Rev. 2006, 36, 267–278.

26. Beach, A.; Gartrell, M.; Han, R. Solutions to Security and Privacy Issues in Mobile Social Networking. In Proceedings of 2009 International Conference on Computational Science and Engineering (CSE), Vancouver, Canada, 29–31 August 2009; IEEE: Los Alamitos, CA, USA, 2009; Volume 4, pp. 1036–1042.

27. Please Rob Me. Available online: http://pleaserobme.com (accessed on 18 February 2010).

28. Twitter: ongoing DoS. Available online: http://status.twitter.com/post/157191978/ongoing-denial-of-service-attack (accessed on 19 January 2010).

29. Whalen, S. An Introduction to ARP Spoofing. Available online: http://packetstorm.securify.com/papers/protocols/intro_to_arp_spoofing.pdf (accessed on 19 January 2010).

30. Malladi, S.; Alves-Foss, J.; Heckendorn, R.B. On Preventing Replay Attacks on Security Protocols. In Proceedings of the International Conference on Security and Management, Las Vegas, NV, USA, June 2002; CSREA Press: Las Vegas, NV, USA, 2002; pp. 77–83.

31. Moss, C.; Evans, D.W. Securing wireless communication against wormhole attacks. In Proceedings of EHAC'05: The 4th WSEAS International Conference on Electronics, Hardware, Wireless and Optical Communications, Salzburg, Austria, 13–15 February 2005; World Scientific and Engineering Academy and Society: Stevens Point, WI, USA, 2005; pp. 1–6.

32. Harbin, J.; Mitchell, P.; Pearce, D. Wireless sensor network wormhole avoidance using disturbance-based routing schemes. In Proceedings of the IEEE 6th International Symposium on Wireless Communication Systems (ISWCS), Siena, Italy, 7–10 September 2009; pp. 76–80.

33. Triki, B.; Rekhis, S.; Boudriga, N. Digital Investigation of Wormhole Attacks in Wireless Sensor Networks. In Proceedings of NCA '09: The 2009 Eighth IEEE International Symposium on Network Computing and Applications, Cambridge, MA, USA, 9–11 July 2009; IEEE Computer Society: Washington, DC, USA, 2009; pp. 179–186.

34. Xu, Y.; Chen, G.; Ford, J.; Makedon, F. Detecting Wormhole Attacks in Wireless Sensor Networks. In Critical Infrastructure Protection; Springer: New York, NY, USA, 2007.

35. Poovendran, R.; Lazos, L. A graph theoretic framework for preventing the wormhole attack in wireless ad hoc networks. Wireless Networking 2007, 13, 27–59.

36. Kan, T.; Kerins, T.; Kursawe, K. Security in Next Generation Consumer Electronic Devices. In Proceedings of Securing Electronic Business Processes–Highlights of the Information Security Solutions Europe 2006 Conference, Rome, Italy, 10–12 October 2006; Vieweg: Wiesbaden, Germany, 2006; pp. 45–53.

37. Saponas, T.S.; Lester, J.; Hartung, C.; Agarwal, S.; Kohno, T. Devices that tell on you: privacy trends in consumer ubiquitous computing. In Proceedings of SS '07: 16th USENIX Security Symposium on USENIX Security Symposium, Boston, MA, USA, 6–10 August 2007; USENIX Association: Berkeley, CA, USA, 2007; pp. 1–16.

38. Dilparic, L.; Arvind, D.K. Design and Evaluation of a Network-Based Asynchronous Architecture for Cryptographic Devices. In Proceedings of ASAP '04: The 15th IEEE International Conference on Application-Specific Systems, Architectures and Processors, Galveston, TX, USA, 27–29 September 2004; IEEE Computer Society: Washington, DC, USA, 2004; pp. 191–201.

39. Srinivasan, V.; Stankovic, J.; Whitehouse, K. Protecting your daily in-home activity information from a wireless snooping attack. In Proceedings of UbiComp '08: The 10th international conference on Ubiquitous computing, Seoul, South Korea, 21–24 September 2008; ACM: New York, NY, USA, 2008; pp. 202–211.

40. Jiang, T.; Wang, H.J.; Hu, Y.C. Preserving location privacy in wireless LANs. In Proceedings of MobiSys '07: The 5th international conference on Mobile systems, applications and services, San Juan, Puerto Rico, 11–14 June 2007; ACM Press: New York, NY, USA, 2007; pp. 246–257.

41. GÃrlach, A.; Heinemann, A.; Terpstra, W.W. Survey on Location Privacy in Pervasive Computing. In Privacy, Security and Trust within the Context of Pervasive Computing, the Kluwer International Series in Engineering and Computer Science; Kluwer (Springer) Academic Publishers: New York, NY, USA, 2004; pp. 23–34.

42. Beresford, A.R.; Stajano, F. Location Privacy in Pervasive Computing. IEEE Pervasive Comput. 2003, 2, 46–55.

43. Karasawa, K.; Kira, Y.; Tsuchiya, Y.; Yamada, K.; Takahashi, K. A Detachable IPsec Device for Secure Consumer Communication Platform. In Proceedings of IEEE Consumer Communications and Networking Conference Poster Session, Las Vegas, NV, USA, 3–6 January 2005.

44. St. Jude's WiFi Pacemaker Wins Approval. Available online: http://dotmed.com/news/story/9878 (accessed on 28 January 2010).

45. Halperin, D.; Heydt-Benjamin, T.S.; Fu, K.; Kohno, T.; Maisel, W.H. Security and Privacy for Implantable Medical Devices. IEEE Pervasive Comput. 2008, 7, 30–39.

46. Denning, T.; Fu, K.; Kohno, T. Absence makes the heart grow fonder: new directions for implantable medical device security. In Proceedings of HOTSEC '08: The 3rd Conference on Hot Topics in Security, Boston, MA, USA, 28 July–1 August 2008; USENIX Association: Berkeley, CA, USA, 2008; pp. 1–7.

47. Drew, T.; Gini, M. Implantable medical devices as agents and part of multiagent systems. In Proceedings of AAMAS '06: The Fifth International Joint Conference on Autonomous Agents and Multiagent Systems, Hakodate, Japan, 8–12 May 2006; ACM: New York, NY, USA, 2006; pp. 1534–1541.

48. Gupta, S.K.S.; Mukherjee, T.; Venkatasubramanian, K. Criticality Aware Access Control Model for Pervasive Applications. In Proceedings of PERCOM '06: The Fourth Annual IEEE International Conference on Pervasive Computing and Communications, Pisa, Italy, 13–17 March 2006; IEEE Computer Society: Washington, DC, USA, 2006; pp. 251–257.

49. Zhang, Y.T.; Bao, S.D.; Shen, L.F. Physiological Signal Based Entity Authentication for Body Area Sensor Networks and Mobile Healthcare Systems. In Proceedings of 27th IEEE Conference on Engineering in Medicine and Biology, Shanghai, China, 1–4 September 2005; pp. 2455–2458.

50. Cherukuri, S.; Venkatasubramanian, K.; Gupta, S. Biosec: A biometric based approach for securing communication in wireless networks of biosensors implanted in the human body. In Proceedings of International Conference on Parallel Processing Workshop, Kaohsiung, Taiwan, 6–9 October 2003; pp. 432–439.

51. Mahimkar, A.; Shmatikov, V. Game-Based Analysis of Denial-of-Service Prevention Protocols. In Proceedings of CSFW '05: The 18th IEEE Workshop on Computer Security Foundations, Aix-en-Provence, France, 20–22 June 2005; IEEE Computer Society: Washington, DC, USA, 2005; pp. 287–301.

52. Andersen, D.G. Mayday: distributed filtering for internet services. In Proceedings of USITS '03: The 4th conference on USENIX Symposium on Internet Technologies and Systems, Seattle, WA, USA, 26–28 March 2003; USENIX Association: Berkeley, CA, USA, 2003; pp. 3–3.

53. Barker, W.C. Recommendation for the Triple Data Encryption Algorithm (TDEA) Block Cipher; NIST Special Publication 800-67; National Institute of Standards and Technology: Gaithersburg, MD, USA, 2008.

54. Rivest, R.; Shamir, A.; Adleman, L. A Method for Obtaining Digital Signatures and Public-Key Cryptosystems. Commun. ACM 1978, 21, 120–126.

55. Desmedt, Y.G.; Frankel, Y. Threshold cryptosystems. In Proceedings of CRYPTO '89: The 9th Annual International Cryptology Conference, Santa Barbara, CA, USA, 20–24 August 1989; Springer-Verlag: New York, NY, USA, 1989; pp. 307–315.

56. Gemmel, P. An introduction to threshold cryptography. CryptoBytes 1997, 2, 7–12.

57. Feng, Y.; Liu, Z.; Li, J. Securing Membership Control in Mobile ad hoc Networks. In Proceedings of ICIT '06: The 9th International Conference on Information Technology, Mumbai, India, 18–21 December 2006; IEEE Computer Society: Washington, DC, USA, 2006; pp. 160–163.

58. Khalili, A.; Katz, J.; Arbaugh, W.A. Toward Secure Key Distribution in Truly Ad-Hoc Networks. In Proceedings of SAINT-W '03: The 2003 Symposium on Applications and the Internet Workshops, Orlando, Florida, USA, 27–31 January 2003; IEEE Computer Society: Washington, DC, USA, 2003; p. 342.

59. Kong, J.; Zerfos, P.; Luo, H.; Lu, S.; Zhang, L. Providing Robust and Ubiquitous Security Support for Mobile ad hoc Networks. In Proceedings of ICNP '01: The Ninth International Conference on Network Protocols, Riverside, CA, USA, 11–14 November 2001; IEEE Computer Society: Washington, DC, USA, 2001; p. 251.

60. Luo, H.; Lu, S. Ubiquitous and Robust Authentication Services for ad hoc Wireless Networks; Technical Report UCLA-CSD-TR-200030; University of California, Los Angeles: Los Angeles, CA, USA, 2000.

61. Lazos, L.; Poovendran, R. SeRLoc: Robust localization for wireless sensor networks. ACM TOSN 2005, 1, 73–100.

62. Lanzisera, S.; Lin, D.T.; Pister, K.S.J. RF time of flight ranging for wireless sensor network localization. In Presented at Workshop on Intelligent Solutions in Embedded Systems (WISES), Vienna, Austria, 30 June 2006; pp. 1–2.

63. Chiang, J.T.; Haas, J.J.; Hu, Y.C. Secure and precise location verification using distance bounding and simultaneous multilateration. In Proceedings of WiSec '09: The Second ACM Conference on Wireless Network Security, Zurich, Switzerland, 16–18 March 2009; ACM: New York, NY, USA, 2009; pp. 181–192.

64. Capkun, S.; Hubaux, J.P. Secure positioning of wireless devices with application to sensor networks. In Proceedings of INFOCOM 2005: 24th Annual Joint Conference of the IEEE Computer and Communications Societies, Miami, FL, USA, 13–17 March 2005; Volume 3, pp. 1917–1928.

65. Sastry, N.; Shankar, U.; Wagner, D. Secure Verification of Location Claims. In Presented at ACM Workshop on Wireless Security (WiSe 2003), San Diego, CA, USA, 19 September 2003; pp. 1–10.

66. Brands, S.; Chaum, D. Distance-bounding protocols. In Proceedings of EUROCRYPT '93: Workshop on the Theory and Application of Cryptographic Techniques on Advances in Cryptology, Lofthus, Norway, 23–27 May 1993; Springer-Verlag: Secaucus, NJ, USA, 1994; pp. 344–359.

67. Kolodziej, K.W.; Hjelm, J. Local Positioning Systems: LBS Applications and Services; CRC Press: Boca Raton, FL, USA, 2006.

68. Wireless LAN Medium Access Control (MAC) and Physical Layer (PHY) Specifications; IEEE 802.11-2007; IEEE Standards Association: Piscataway, NJ, USA, 2007.

69. Ekici, E.; Vural, S.; McNair, J.; Al-Abri, D. Secure probabilistic location verification in randomly deployed wireless sensor networks. Ad Hoc Networks 2008, 6, 195–209.

70. Liu, Y.; Zhou, H.; Zhao, B. Secure Location Verification Using Hop-Distance Relationship in Wireless Sensor Networks. In Proceedings of APSCC '07: The 2nd IEEE Asia-Pacific Service Computing Conference, Tsukuba Science City, Japan, 12–14 December 2007; IEEE Computer Society: Washington, DC, USA, 2007; pp. 62–68.

71. Vural, S.; Ekici, E. Analysis of hop-distance relationship in spatially random sensor networks. In Proceedings of MobiHoc '05: The 6th ACM

International Symposium on Mobile ad hoc Networking and Computing, Urbana-Champaign, IL, USA; ACM: New York, NY, USA, 2005; pp. 320–331.

72. Wei, Y.; Yu, Z.; Guan, Y. Location Verification Algorithms for Wireless Sensor Networks. In Proceedings of the 27th International Conference on Distributed Computing Systems (ICDCS), Toronto, Ontario, Canada, 25–29 June 2007; pp. 70–70.

73. Leinmuller, T.; Schoch, E.; Kargl, F. Position Verification Approaches for Vehicular ad hoc Networks. IEEE Wirel. Commun. 2006, 13, 16–21.

74. Zhang, Y.; Liu, W.; Fang, Y.; Wu, D. Secure localization and authentication in ultra-wideband sensor networks. IEEE J. Sel. Area. Commun. 2006, 24, 829–835.

75. Hightower, J.; Borriello, G. A Survey and Taxonomy of Location Sensing Systems for Ubiquitous Computing; UW CSE 01-08-03; University of Washington: Seattle, WA, USA, 2001.

76. Srinivasan, A.; Wu, J. A Survey on Secure Localization in Wireless Sensor Networks. In Encyclopedia of Wireless and Mobile Communications; CRC Press, Taylor and Francis Group: Boca Raton, FL, USA, 2008.

77. Han, C.; Zhan, S.; Yang, Y. Proactive attacker localization in wireless LAN. SIGCOMM Comput. Commun. Rev. 2009, 39, 27–33.

CHAPTER 6

Mobile M2M communication architectures, upcoming challenges, applications, and future directions

Yasir Mehmood [1*], Carmelita Görg[1], Maciej Muehleisen[2] and Andreas Timm-Giel[2]

[1]Communication Networks, University of Bremen, Bremen, Germany.

[2]Institute of Communication Networks, University of Technology, Hamburg, Germany.

ABSTRACT

Machine-to-machine (M2M) communication is becoming an increasingly important part of mobile traffic and thus also a topic of major interest for mobile communication research and telecommunication standardization bodies. M2M communication offers various ubiquitous services and is one of the main enablers of the vision inspired by the Internet of Things (IoT). The concept of mobile M2M communication has emerged due to the wide range, coverage provisioning, high reliability, and decreasing costs of future mobile networks. Nevertheless, M2M communications pose significant challenges to mobile networks, e.g., due to the expected large number of devices with simultaneous access for sending small-sized data, and a diverse application range. This paper provides a detailed survey of M2M communications in the context of mobile networks, and thus focuses on the latest Long-Term Evolution-Advanced (LTE-A) networks. Moreover, the end-to-end network architectures and reference models for M2M communication are presented. Furthermore, a comprehensive survey is given to M2M service requirements, major current standardization efforts, and upcoming M2M-related challenges. In addition, an overview of upcoming M2M services expected in 5G networks is presented. In the end, various mobile M2M applications are discussed followed by open research questions and directions.

KEYWORDS

Machine-to-machine Internet of Things End-to-end Long-term evolution-advanced Standardization efforts 5G

1 REVIEW

1.1 Introduction

In recent years, collaborative communication between numerous intelligent systems through either mobile or fixed networks has achieved significant importance. One of the resultant emerging domains is mobile machine-to-machine (M2M) communication. M2M communication describes a communication style in which two or more entities such as devices/machines communicate with each other autonomously [1]. M2M communication is playing a promising role in enabling the Internet of Things (IoT) vision by providing ubiquitous connectivity between numerous intelligent devices [2]. The term M2M in IoT describes the autonomous exchange of information among numerous devices inter-connected with each other. M2M communication thus covers a wide range of use cases [3–5]. For instance, the M2M application area includes intelligent transportation systems (ITS), logistics and supply chain management, smart metering, e-healthcare, surveillance and security, smart cities, and home automation. Particularly, many intelligent devices are expected to be deployed in the automotive sector, security, e-healthcare, and logistics [6]. Consequently, M2M communications are anticipated to reshape the business/revenue of telecom operators, M2M enterprises, and M2M enablers due to the emerging use of intelligent sensors and actuators in many advanced aforementioned applications [7].

Mobile standards are foreseen to play a prominent role in the successful deployment of numerous M2M applications due to several reasons, e.g., high capacity to support a large number of devices, enhanced coverage, low power consumption, high reliability, low cost modules, as well as easy and short-term deployment opportunities [2]. Several forecasts have reported a considerable market growth for M2M device volume. Some of these recent forecasts predicting the number of cellular-based M2M devices worldwide are shown in Fig. 1. Therefore, M2M services are considered as one of the main motivations behind the tremendous growth in mobile traffic [6]. Furthermore, mobile networks such as Long-Term Evolution/Long-Term Evolution-Advanced

(LTE/LTE-A) offer high data rates, enhanced coverage, and improved radio resource management mechanisms. However, the cellular systems are particularly designed to support the exponentially growing human-to-human/human-to-machine (H2H/H2M) traffic by enabling improved broadband services [8]. For instance, Ericsson in its mobility report [9] anticipates approximately 5.6 billion cellular-based active smartphones by 2019. In order to support increasing mobile traffic, several enhancements in existing LTE systems [10] such as femtocells [11] and relay nodes (RNs) [12] were introduced to fulfill the International Mobile Telecommunications-Advanced (IMT-Advanced) demands of maximum throughput of up to 1 Gbps in downlink and 500 Mbps in uplink [13].

Figure 1 Recent forecasts about cellular-based M2M traffic volume by IDC [141], M2M magazine [88], CISCO [85], GSMA [87], and ABI research [6]

Several review papers [14–22] discussed M2M communications in the context of wireless technologies by focusing on various M2M applications such as e-healthcare, smart metering, smart grids, IoT paradigm, and radio resource management. For instance, authors in [14] addressed the M2M e-healthcare application, i.e., m-Health, by highlighting the benefits of using M2M technology for both patients and service providers due to reliable autonomous communication. Furthermore, in [14], a first detailed analysis of wireless body area networks (WBANs) towards the patient side is provided, and various end-to-end (E2E) solutions for practical e-healthcare applications were discussed. In [20], the authors surveyed a number of existing communication technologies which can be used for M2M communication in smart grids. Furthermore, the authors presented the most suitable technologies such as Institute of Electrical and Electronics Engineering *(IEEE) 802.15.3a* (Ultra Wide Band), *IEEE 802.15.1* (Bluetooth), and *IEEE 802.15.4* (ZigBee) for smart grids by addressing their limitations as well. Authors in [22] surveyed existing Internet Engineering Task Force (IETF) protocols for supporting the IoT paradigm. In [23], authors discussed applicability of software-defined networks and network virtualization to mobile and wireless networks. Additionally, authors in [15] highlighted the correlation between M2M, wireless sensors networks (WSNs), cyber-physical systems (CPS), and the IoT. Furthermore, authors in [15] proposed an M2M system integrating intelligent roads with unmanned vehicles. In [16], authors surveyed several M2M communication scenarios in the context of wireless technology to cloud and a machine swarm of a large number of devices and explored their practical realizations. In [17, 18], authors presented various radio resource management and scheduling schemes for mobile M2M communications. Moreover, issues regarding random access due to the large number of devices using mobile networks are addressed, e.g., in [19] and [21].

To the best of the authors' knowledge, a comprehensive survey covering existing mobile M2M communication architectures, service requirements, existing and upcoming challenges, up-to-date standardization efforts, application scenarios, upcoming M2M services in 5G networks, as well as future directions is not available. Therefore, the objective of this survey is to give a comprehensive overview on the aforementioned topics which helps the reader to understand the concept, future trends, and requirements of mobile M2M communication. Since LTE and LTE-A are the latest Third Generation Partnership Project (3GPP) mobile standards, the rest of the paper focuses on M2M communications in LTE/LTE-A networks. Furthermore, an overview of the rest of the survey and a list of most frequently used acronyms is presented in Fig. 2 and Table 1, respectively.

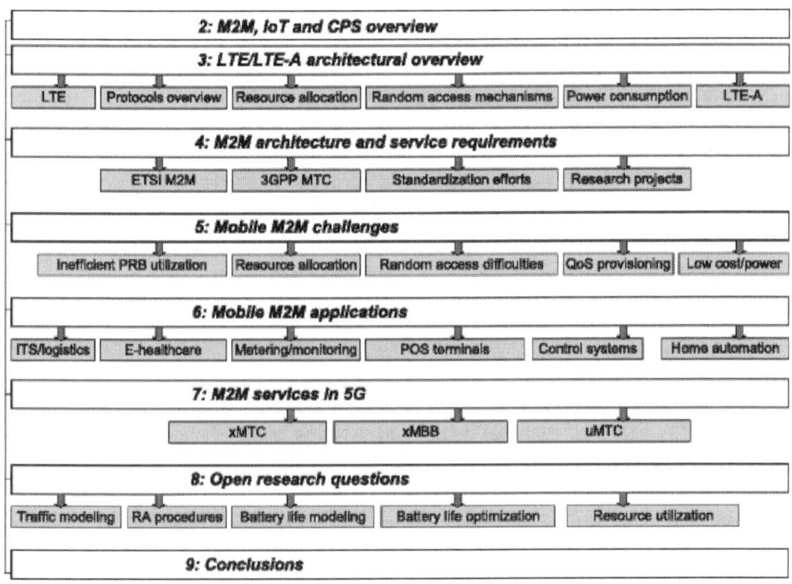

Figure 2 An overview of the survey

Table 1 List of used abbreviations

Abbreviations	Acronyms
AAA	Accounting, authorization, and authentication
ARQ	Automatic repeat request
AS	Access stratum
CP	Cyclic prefix
CMPP	Coupled Markov modulated Poisson process
IMT-Advanced	International Mobile Telecommunications-Advanced
EPS	Evolved packet system
E-UTRAN	Evolved UMTS terrestrial radio network
GERAN	GSM EDGE radio access network
GGSN	Gateway GPRS support node
GSM	Global system for mobile communications
HARQ	Hybrid automatic repeat request
HLR	Home location register
HSDPA	High-speed downlink packet access
HSS	Home subscriber server
HSUPA	High-speed uplink packet access
ICT	Information and communication technology
ITU	International Telecommunication Union
MAC	Medium access control
MME	Mobility Management Entity
MMPP	Markov modulated Poisson process
MSC	Mobile switching center
MTC-IWF	MTC InterWorking Function
NAS	Non access stratum
OMA	Open mobile alliance
PDNs	Packet data networks
P-GW	Packet gateway
PHY	Physical layer
PLMN	Public land mobile network
PSM	Power saving mode
PRB	Physical resource block
QCI	QoS class identifiers
RLC	Radio link control
RRC	Radio resource control
SAE	System architecture evolution
SCS	Services capability server
S-GWs	Serving gateways
SGSN	Serving GPRS support node
SMS-SC	Short message service-service center
TTI	Transmission time interval
UMTS	Universal mobile telecommunications system
WLAN	Wireless local area network
WPAN	Wireless personal area network
WSN	Wireless sensor network

1.2 M2M, IoT, and CPS: an overview

The use of communication technologies for ubiquitous connectivity among numerous devices/machines has achieved significant importance. As a result, notable developments have been made in several domains such as M2M, WSNs, IoT, and CPS. Since the key idea in the above domains is identical, the authors discussed correlations among M2M, WSNs, IoT, and CPS in [24]. M2M communication offers autonomous communication between intelligent sensors and actuators. Therefore, M2M traffic characteristics differ from conventional mobile traffic, e.g., in terms of traffic density, packet size, and quality of service (QoS) requirements [25]. For instance, Fig. 3 presents some of the data sizes of various M2M uses cases. Additionally, an experimental study done in [26] shows that M2M traffic exhibits a significantly different behavior than the traditional smartphone (mobile) traffic in various aspects. For example, unlike traditional mobile traffic which is dominant in downlink rather than in uplink, M2M is an uplink-dominant traffic which particularly generates bursty traffic volumes. Furthermore, M2M devices compete with smartphones for network resources in co-geographical regions. Thus, supporting a large number of devices and issues such as spectral efficiency enhancements in the presence of M2M traffic, latency control, and limitation of network congestion demands utmost attention. These findings suggest that more careful spectrum allocation and utilization schemes are needed, so that M2M communications can benefit from mobile networks such as LTE/LTE-A without degrading the performance of their traditional traffic. For these reasons, well-known standardization bodies such as 3GPP and European Telecommunication Standardization Institute (ETSI) have already started putting efforts to support a large number of devices while considering their characteristics and service requirements. In addition, 3GPP and ETSI have already proposed an E2E reference model and functional architectures, respectively [27, 28]. Moreover, 3GPP has considered to investigate MTC[1] service requirements and subsequently proposed various schemes for system improvements in the presence of MTC.

The term IoT describes connecting objects/machines often called *things* with an infrastructure based on the Internet. The fundamental components of IoT include sensing/acquiring data, heterogeneous connectivity, data processing, as well as applications and services. Additionally, security and privacy are also considered as essential components in IoT. Moreover, in IoT, WSNs are usually considered as a fundamental scenario which spatially distributes autonomous sensors for monitoring various physical and environmental factors, thus to deliver the acquired information to backend stations. IoT covers a vast application area such as greenhouse monitoring, ITS, smart metering/monitoring, smart lighting, smart grids, and telemedicine monitoring.

Several applications particularly related to intelligent and autonomous control are not yet considered as IoT-based applications [24]. However, developments in the domains of cloud computing and distributed multi-agent control along with WSNs and IoT have evolved M2M into a new form of communication called CPS. CPS has emerged as a prominent technology which is capable to support M2M and non-M2M applications. CPS offers communication and coordination among physical elements of a system and focuses on, e.g., interactive applications as well as distributed real-time control. CPS is an emerging form of IoT and can significantly benefit from communication technologies to provide services on the basis of information acquired from the surroundings (physical world). Figure 4 depicts the relationships and correlations among M2M, WSNs, IoT, and CPS [8]. The IoT is represented by the space formed by M2M, WSNs, and CPS. Additionally, Fig. 4 shows that developments in M2M and WSNs are expected to bring forth immense CPS applications in the future. Furthermore, authors concluded that currently M2M is considered as one of the crucial parts of IoT, whereas CPS is an evolution of M2M technology. Consequently, CPS will be regarded as one of the major technical frameworks for IoT.

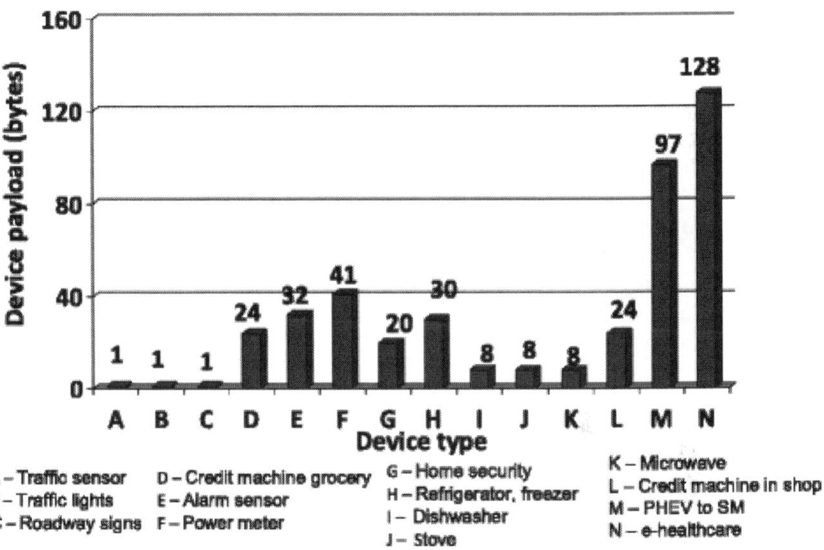

Figure 3 Traffic models of various M2M applications representing small-sized data packets [142]

1.3 LTE/LTE-A architectural overview

This section presents an overview of the LTE/LTE-A network architecture and protocols. Since this survey focuses on mobile M2M communication, an overview of the relevant LTE topics such as radio resource management, random access procedures, and UE power consumption is presented. Later on, these topics are thoroughly discussed in the context of mobile M2M communication challenges in Section 1.5.

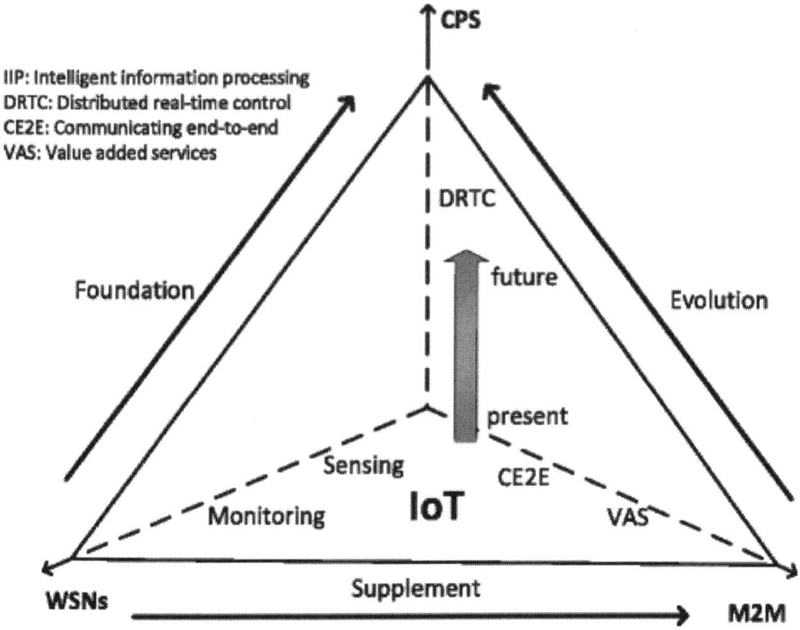

Figure 4 A representation of correlations among M2M, WSNs, IoT, and CPS. Moreover, it shows that advancements in M2M and WSNs are expected to increase the number of CPS applications in the future (re-drawn from [15])

1.3.1 Long-Term Evolution

Unlike traditional mobile networks such as global system for mobile communications (GSM), LTE is the first fully packet-switched mobile network proposed by 3GPP in 2004. 3GPP standardized LTE in its Release 8 with small modifications in Release 9 [13, 29]. The standard is designed on the basis of the GSM and universal mobile telecommunications system (UMTS) networks to provide improved energy and spectral efficiency. The LTE radio and non-radio enhancements include evolved UMTS terrestrial radio network (E-UTRAN) and system architecture evolution (SAE), respectively. The evolved packet system (EPS) is the combination of LTE and SAE. The function of the EPS bearer[2] is to route traffic from gateways (GWs) in the packet data networks (PDNs) to user equipments (UEs). A high level E-UTRAN architecture is presented in Fig. 5 (right). According to the 3GPP Technical Report TR 25.913 [30], the details of the performance requirements of 3GPP LTE networks can be described as follows:

1 LTE supports a maximum data rate of 100 and 50 Mbps in downlink and uplink, respectively (for a 20-MHz bandwidth).

2 In downlink, the LTE mean throughput for a user is three to four times that of high-speed downlink packet access (HSDPA) (Release 6) per megahertz.

In uplink, the average user throughput is two to three times that of high-speed uplink packet access (HSUPA) (Release 6) per megahertz.

3 The LTE control plane capacity supports approximately 200 active UEs per cell for a 5-MHz bandwidth assignment.

4 4.The spectrum efficiency of LTE networks is three to four times that of HSDPA in downlink, whereas in uplink, it is two to three times that of HSUPA (Release 6).

5 Mobility support in LTE is 0–15 km/h (low-speed users) and high-speed mobility support from 15 km/h up to 500 km/h.

6 LTE supports a cell coverage of up to 5 km in order to meet throughput, spectral efficiency, and mobility requirements. The cell coverage can be extended up to 30 km at the cost of minor performance degradation.

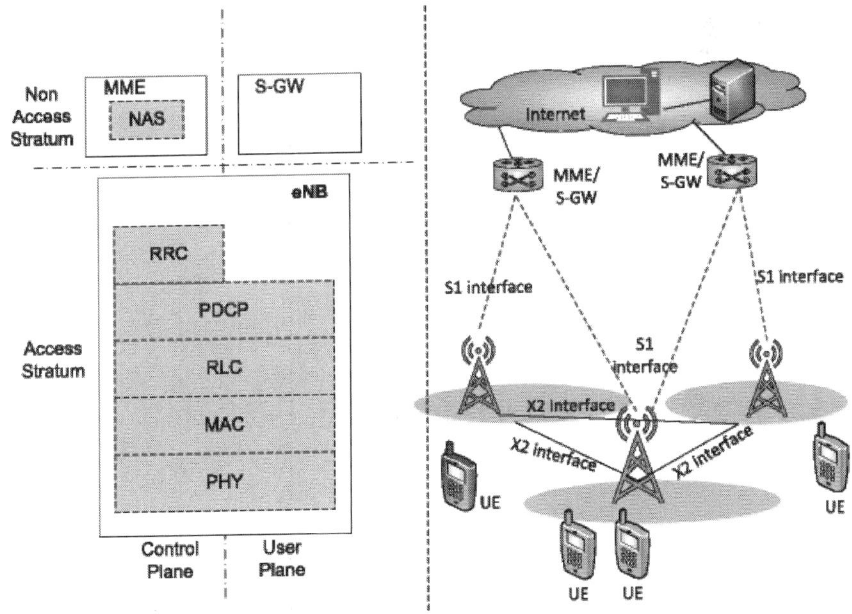

Figure 5 LTE/LTE-A network architecture (*right*) with protocols stack (*left*) [13]

1.3.2 LTE protocol overview

LTE interfaces between UE, E-UTRAN, EPC, and S-GWs are associated with protocol stacks which are used to exchange data and signaling messages [31]. Therefore, the LTE protocol stack can be divided into the following categories:

1 *User plane protocols:* These protocols support routing of users data between UEs and S-GWs.

2 *Control plane protocols:* These protocols are used for exchanging signaling messages between various devices within the network.

On the air interface (Uu), the UE high-level functionalities are controlled by the MME [32, 33]. However, there is no direct communication path between UE and MME. The communication path between UE and MME is established using the evolved node base stations (E-UTRAN eNB) which enhance the level of the hardware complexity in the network. To reduce this complexity, the Uu interface is further divided into two levels of protocols. One is called access stratum (AS) whereas the second is the non-access stratum (NAS). The MME high-level signaling lies in the NAS level but is transported within the network using AS protocols. The control and user plane termination protocols are supported by the eNB. The user plane protocols include the packet data convergence protocol (PDCP), the medium access control (MAC), and the physical (PHY) layer protocols. In addition, the protocols of the control plane include the radio resource control (RRC) protocols. A high level overview of the protocol stack is depicted in Fig. 5 (left). The major functionalities of these protocols are given below:

1 Access stratum (AS) [34]: UEs are capable to access the capabilities and services of communication networks through the AS protocols. There are several functionalities of the radio AS protocols such as dynamic allocation of radio resources to UEs (UL/DL), controlling bearers, traffic management, radio admission control, and handover management.

2 NAS: The NAS functionalities include the establishment of radio connections between the network and the UE. Moreover, registration, authentication, local registration management, and mobility management are also included in the functionalities of the NAS protocols.

3 RRC [35]: The major RRC functions and services include broadcasting system information, establishing/releasing RRC connections, security management, mobility functions, paging, QoS management, and direct messaging between UE and NAS.

4 PDCP [36]: The PDCP layer provides several services such as transfer of user and control plane data, header compression, re-transmissions of SDUs lost during handover, ciphering, detection of duplicate data, and in-sequence delivery.

5 Radio link control (RLC) [37]: There are several functionalities of the RLC layer such as the segmentation of data packets according to the available size of the transport block. Moreover, the error correction through automatic repeat request (ARQ) and the re-segmenting RLC data PDUs are included in the functionalities of the RLC. In addition, RLC performs functions such as concatenation, segmentation as well as reassembly of RLC SDUs, packet delivery, and error detection in protocols.

6 MAC [38]: The MAC layer functionalities include mapping between logical and control channels, RLC SDUs multiplexing/demultiplexing, scheduling, error corrections using hybrid ARQ (HARQ), priority handling of local channels, segmentation and reassembly of upper layer PDUs, padding, and ciphering.

7 PHY [39]: The PHY layer provides data transport services to the upper layers such as synchronization of time and frequency, physical channel modulation/demodulation, encoding/decoding of transport channels, MIMO operations, and transmit diversity. The frequency and time variation in LTE are exploited through orthogonal frequency-division multiplexing (OFDM) and single carrier frequency division multiple access (SC-FDMA) techniques. The physical layer sub-carrier spacing in LTE is 15 kHz.

1.3.3 LTE radio resource management

LTE transmission schemes support orthogonal frequency division multiple access (OFDMA) and SC-FDMA in downlink and uplink, respectively. The radio frame structure in the LTE uplink and downlink is similar. In LTE evolved universal terrestrial radio access (E-UTRA), the channel bandwidth ranges from 1 to 20 MHz [40]. The E-UTRA channel bandwidths include 1.4, 3, 5, 10, 15, and 20 MHz. Each LTE frame (10 ms) is divided into ten subframes of 1 ms each [34], see Fig. 6. Every single subframe consists of two equal time slots of 0.5 ms. Each slot is further divided into six and seven OFDM symbols in the extended and the normal cyclic prefix (CP), respectively. For uplink and downlink communications in LTE, the smallest radio resource which can be allocated to each UE is called PRB which has a dimension of 180 KHz in the frequency domain and 1 ms in the time domain. A single PRB is further composed of 12 consecutive sub-carriers. In normal cyclic prefix, there are 84 resource elements per PRB whereas an extended cyclic prefix is composed of 72 resource elements per PRB.

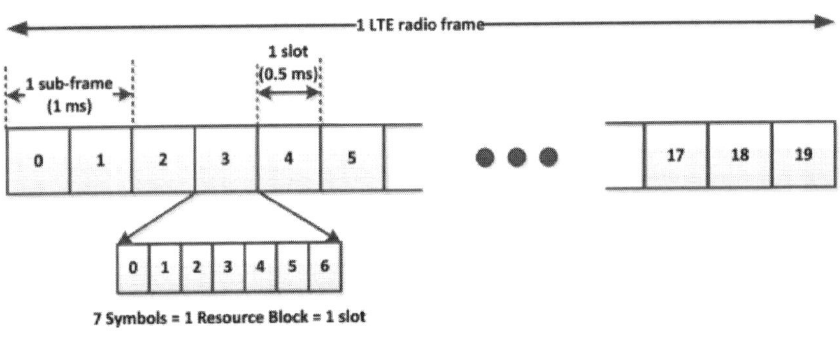

Figure 6 LTE FDD frame structure [13]

LTE radio resource management (RRM) enables efficient utilization of the scarce radio resources among several UEs. RRM consists of a set of procedures to distribute available PRBs among multiple users while ensuring their QoS as well as seamless connectivity [41]. The scheduling of radio resources in LTE is performed by the MAC layer of the eNB. Downlink and uplink scheduling procedures are initiated in every transmission time interval (TTI) of 1 ms depending on the amount of data in the buffers of eNB and UE, respectively. Furthermore, the RRM initiates the scheduling process when sufficient PRBs are available in the network to meet the QoS demands of other users as well. The transport block size (TBS) represents the capacity to transmit data with respect to the RRM procedure. In addition, the capacity of TBS depends on the available modulation and coding scheme (MCS) as well as the number of allocated PRBs to a particular UE. Figure 7 represents the capacity of TBS (in bits) with respect to the MCS index and the number of allocated PRBs. It can be seen from Fig. 7 that TBS capacity increases for higher values of MCS and larger number of allocated PRBs. For instance, 328 bits can be sent per PRB with an MCS index of 16. However, the PRB capacity increases to 712 bits when the index of MCS is raised to 26.

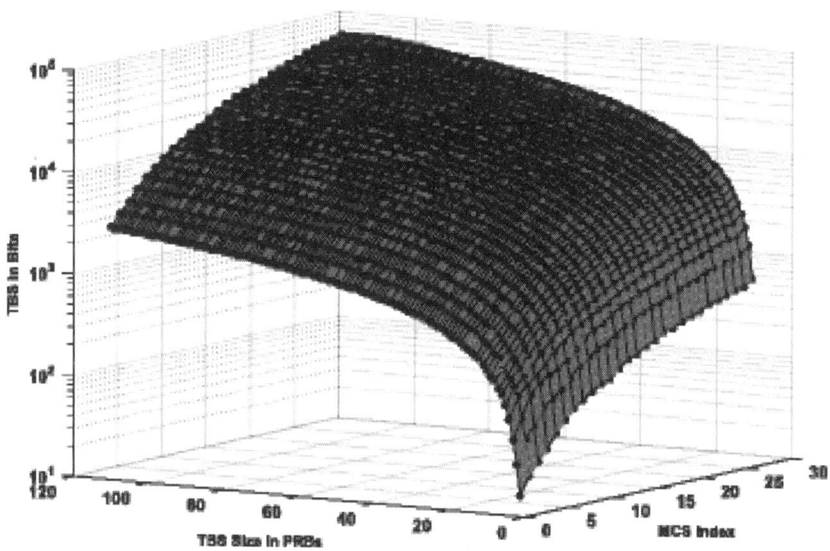

Figure 7. An illustration of LTE TBS in bits with respect to MCS index and the number of allocated PRBs

1.3.4 LTE random access mechanisms

In LTE networks, each UE performs random access (RA) procedures to transmit data during the access grant time interval (AGTI) or RA slots over a time-frequency radio resource called physical random access channel (PRACH). A UE initiates the RA mechanism in two cases [42]. Firstly, when a UE is turned on and the uplink timing synchronization is either lost or not yet achieved. Secondly, whenever a handover mechanism is performed from a serving eNB to a target eNB. In LTE, UEs perform the RA procedure for several reasons such as for establishing a communication link, re-initiating a radio link after a failure, for radio resources in case of no uplink resource grant, and scheduling requests in the absence of dedicated physical uplink control channels (PUCCHs).

The RA procedure can be classified into a *contention-based* and a *contention-free* random procedure [19, 42–44]. Initially, a UE uses a contention-based RA procedure to select a preamble. A preamble is an OFDM-based signal with a narrower sub-carrier spacing. It is possible that several devices can select the same preamble, which demands a further contention resolution mechanism. In a contention-free RA procedure, an eNB uses dedicated preambles to initiate the RA mechanism. Hence, it ensures faster response than the contention-based approach. Contention-free RA procedures are usually used during critical situations such as handover. The major steps of contention-based and contention-free RA procedures are given in Fig. 8. Furthermore, a contention-free RA procedure is simpler than the contention-based RA procedure and consists of three steps by allocating dedicated preamble signatures to a device (see Fig. 8). The procedure starts with the RA preamble assignment by the eNB. After the transmission of an assigned RA preamble by the UE, the eNB responds with the random access response (RAR). This is the last step of the contention-free RA since there is no need to resolve further collision.

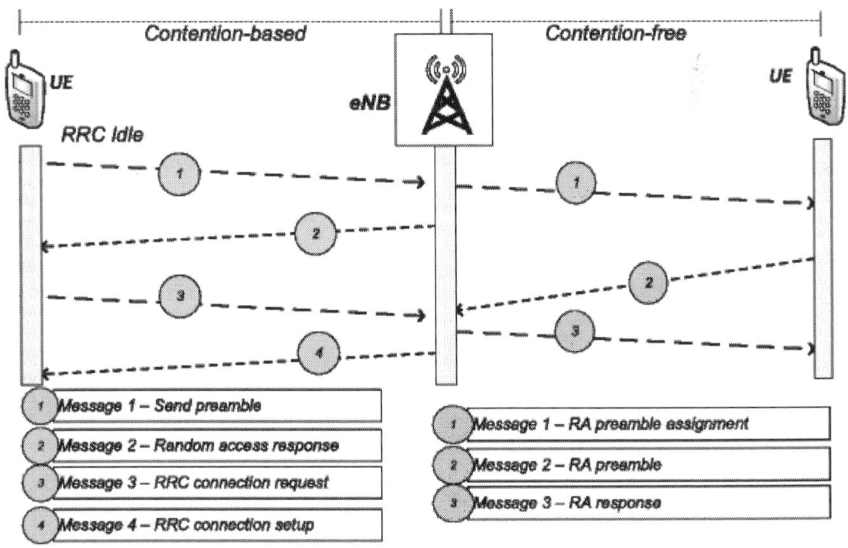

Figure 8. An illustration of contention-based and contention-free RA procedures

1.3.5 UE power consumption in LTE

Battery life optimization for mobile UEs, especially smartphone users, has always remained a challenging task for communication researchers. In the case of smartphones, users charge their phones frequently, usually every day. However, they always demand for the optimum power consumption of phones in order to increase its battery lifetime. In order to fulfill the increasing demand of improved UE power consumption, 3GPP introduced the power-saving mode (PSM) such as discontinuous reception (DRX) and idle state in its Release 12 TS 23.682 [45]. Under the PSM, a UE requests the network for an active timer duration during tracking area updates (TAUs) procedures. In this time, a UE remains attached, however all the AS layer functions are stopped. The DL data can be received by a UE in the RRC connected state, and during the active time period. The active timer is initiated by the UE when it moves from the connected to the idle state. The device moves to PSM with the expiry of the timer. The value of the active timer represents the time in which the device is accessible. In PSM, a device remains attached/registered with the network. However, it is not accessible due to the fact that it does not check for paging in PSM. Moreover, the device stays in PSM until it starts an activity such as UL data transmission. Figure 9 depicts the state transitions of a UE from connected to idle, and vice versa.

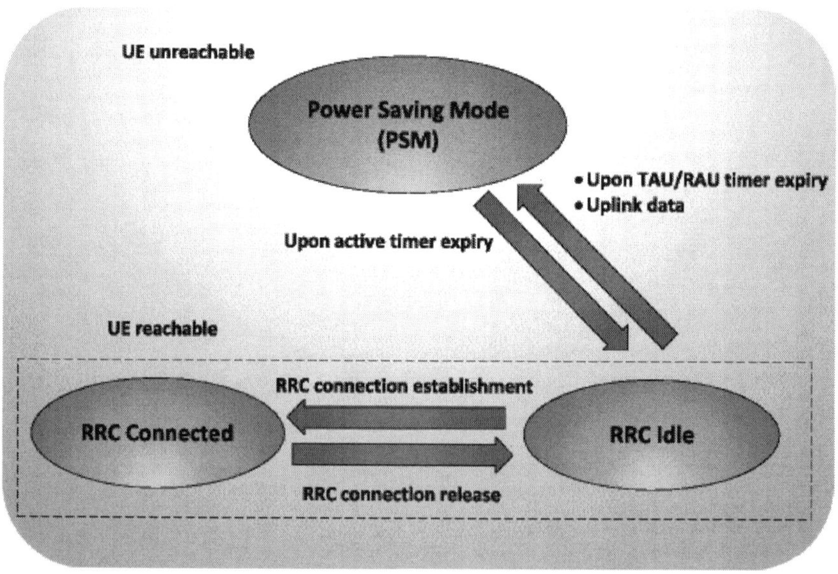

Figure 9.A graphical representation of 3GPP standardized PSM

1.3.6 LTE-Advanced

LTE-A is the latest 3GPP standard which defines further advancements with respect to the existing LTE standard to approach the needs of IMT-Advanced which has been set by the International Telecommunication Union Radio communication (ITU-R) [46]. According to these requirements, LTE-A is capable to provide throughput of up to 1 Gbps in downlink and 500 Mbps in uplink. The peak spectrum efficiency in LTE-A is also improved by 15 Bps/Hz in uplink and 30 Bps/Hz in downlink with the maximum antenna configurations of 4×4 and 8×8, respectively. Moreover, the LTE-A standard is also designed to reduce the latency in user and control planes compared to the existing LTE standard. To achieve the defined tasks for LTE-A, the problems such as low signal-to-interference-plus-noise ratio (SINR) and network coverage issues due to shadowing and non-line-of-sight (NLOS) are considered. According to [47], the capacity of the 3GPP LTE Release 8 network is already very close to the Shannon limit and it is required to upgrade the existing technologies with new standards to support the increasing future data traffic. One idea is to increase the number of eNBs which is not a cost-efficient approach and not an appealing option for network operators.

In LTE-A networks, the evolved UMTS terrestrial radio network (E-UTRAN) consists of eNBs with a similar role as that of the base station (BS) in GSM but with additional functionalities [41, 48]. The E-UTRAN handles communication between the UE and the EPC, whereas an eNB handles a UE in one or more cells. The serving eNB (SeNB) is the eNB which currently provides services to a UE. Moreover, the eNB controls low-level operations such as handover functions. The eNB communicates with the core network through the S1 interface. Several eNBs are also inter-connected with each other through the X2 interface (see Fig. 5, right) which is mainly used for exchanging information such as packet forwarding during handover. The S1 interface between an eNB and EPC can also perform handover functions at the cost of more delay. The S1 and X2 interfaces operate using an IP-based transport network. Moreover, E-UTRAN also provides home eNB (HeNB) services. HeNB is an eNB which is particularly dedicated to a user within a home for femtocell coverage. It can be connected either directly to the EPC in the same way as a standard eNB is connected or using some intermediate terminals such as gateways that forward information from multiple eNBs.

1.4 M2M network architectures

Besides the already mentioned characteristics of M2M traffic, it also exhibits unique characteristics such as an increasing M2M device volume, sending small payloads, the demand for various mobility profiles, time-controlled, and mainly delay-tolerant as well as bursty traffic [26]. Therefore, it is of significant importance to optimize the existing mobile standards which have been originally standardized for traditional mobile traffic. Due to the abovementioned unique characteristics, M2M communications demand an optimized E2E network architecture, particularly over mobile networks such as LTE/LTE-A. To enable global connectivity among various M2M applications, several standardization bodies are working on designing new and optimized architectures to support M2M communications over mobile networks. ETSI has designed and standardized an E2E network architecture for M2M communications by considering various M2M service requirements. Similarly, 3GPP has also defined a reference model for MTC. In this section, both ETSI and 3GPP M2M architecture and reference models are discussed for the understanding of the reader.

1.4.1 ETSI M2M architecture

According to ETSI specifications discussed in [15, 49, 50], the M2M architecture consists of the following five key elements which are shown in Fig. 10:

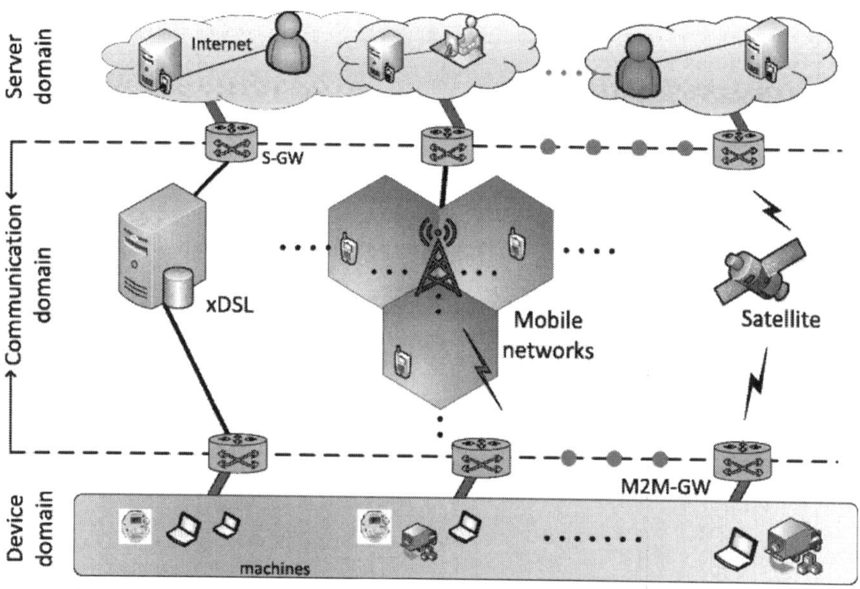

Figure 10 A high-level E2E network architecture for M2M communications proposed by ETSI (re-drawn from [15])

M2M device domain The M2M devices are responsible for the collection and autonomous transmission of sensor data such as the internal temperature and humidity level of a container in logistics, and ITS. The M2M devices are usually connected to small local networks called subnets for the transmission or reception of data to or from the M2M application domains (server domain).

M2M area networks The M2M area networks are responsible for establishing the communication path between the M2M devices and the M2M gateways. These networks are usually called subnets which collect and route information from the M2M devices to the M2M gateways. There are several subnets which are used for generating the communication link between the M2M devices and the M2M gateways [50]. Generally, the use of subnets is dependent on the network technology. In *fully distributed networks*, all M2M devices are connected as peers to the network. One of the nodes which is connected to the network, e.g., via wired or wireless connectivity, acts as a router. *In client-server networks*, all nodes or devices communicate directly with the server. Whereas in *cooperative networks*, all nodes communicate with each other using some intermediate gateways. Some of the major subnets are presented in Table 2.

Table 2 List of technologies for M2M communication [15]

Sr	M2M area networks
1	Power line communication (PLC)
2	Short-range device (SRD)
3	Ultra wide band (UWB)
4	ZigBee
5	Meter bus (M-BUS)
6	Wireless meter bus
7	Bluetooth

M2M gateways The intelligent sensors, which are deployed to collect information, communicate with the communication network with the help of M2M gateways. Hence, M2M gateways act like bridges between the sensors and the communication network.

Communication networks The main role of the communication network is to create a communication path between the M2M devices and application servers either through wired or wireless communication networks. The main communication networks are listed in Table 3.

Server/application domain The M2M application domain consists of a middleware layer where the collected packets pass through several application services and then are used by the related agencies.

Table 3 List of technologies for M2M communications [15]

Sr	M2M communications networks
1	Short-range devices (SRDs)
2	Ultra wide band (UWB)
3	Satellite
4	GSM, UMTS, LTE, LTE-A
5	Wireless LAN
6	WiMAX
7	xDSL

1.4.2 3GPP MTC reference model

In the 3GPP MTC reference model given in [45, 51–53], the communication between MTC devices and application servers use 3GPP services and optionally services capability server (SCS)[3] services. The application server (AS) particularly hosts an MTC application in the external network. Furthermore, the AS manipulates an SCS for supplementary value-added services. The 3GPP provides several services to support MTC, e.g., transport, user management as well as architectural enhancements, e.g., initializing control plane functionalities of an MTC device [52]. 3GPP envisages various models for MTC communications [52]. These models are related to the SCS provider-based communication between the AS and the 3GPP system. The various architectural models proposed by 3GPP include the direct model, the indirect model, and the hybrid model [45, 52, 53]. In the case of the *direct* model, the AS directly performs data communication with the UE by connecting directly to the network and does not use any SCS service (see Fig. 11). In addition, the 3GPP services can be used by the application in the outer network. In the case of the *indirect* model, the AS uses SCS services and connects indirectly to the network. Thus, the indirect model utilizes supplementary value-added services for MTC such as control plane device triggering. In addition, the SCS is either controlled by the MTC service providers or network operators in order to provide several services, e.g., value-added services and communication services related to control and user plane. In case of the hybrid model, both the aforementioned models (direct and indirect) are used by the AS concurrently. The AS performs direct data communications with the device by connecting directly to the network. However, it also uses SCS services. From the 3GPP network aspect, the AS direct data communication and the SCS value-added services, e.g., control plane communication are fully independent, although both can service the same AS MTC application. In addition, the MTC applications can receive distinct services/facilities from the SCS controlled by the MTC service provider as well as the 3GPP operator. Since the abovementioned three models are complementary, 3GPP operators can combine them for various applications. For instance, the SCS based on the MTC service provider can be combined with the 3GPP network operator controlled SCS, both communicating with the similar PLMN.

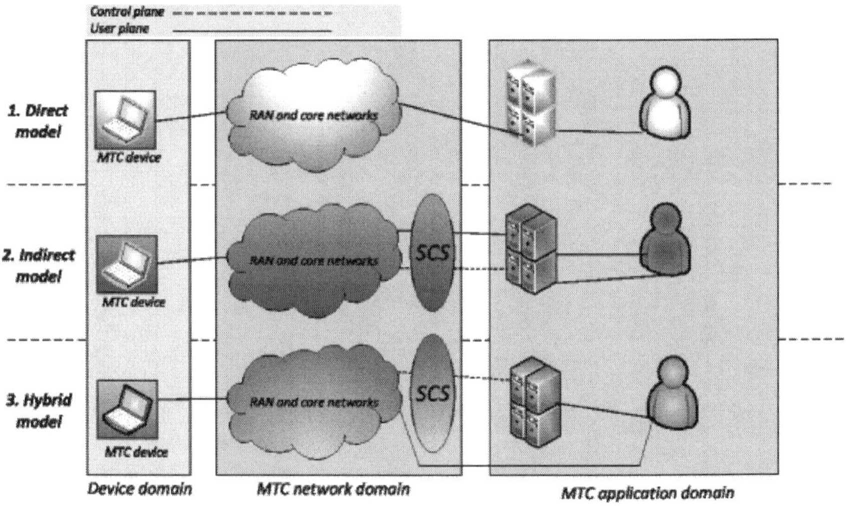

Figure 11. 3GPP direct, indirect, and hybrid models for MTC communications

Architectural reference model The architectural reference model for MTC is depicted in Fig. 12, which shows an MTC device communicating with 3GPP networks, i.e., E-UTRA, E-UTRAN, and GERAN using Um, Uu, and LTE-Uu interfaces, respectively. Moreover, Tsp is used for 3GPP network services to SCS and AS (see Fig. 12). 3GPP has standardized the Tsp interface in order to support value-added services for MTC such as control plane device triggering. In addition, an SCS usually provides Tsp interface. Moreover, the MTC interworking function (MTC-IWF) is a functional entity used to support indirect and hybrid models by residing in the home public land mobile network (HPLMN). It usually shows the internal PLMN network topology. Furthermore, it also translates Tsp-related signaling protocols to bring particular functionality in the PLMN. The descriptions of various MTC-related reference points and network elements are summarized in Table 4 (taken from [53]).

1.4.3 MTC service requirements in mobile networks

The service requirements vary significantly due to the diversity of MTC applications. For this reason, the authors of [15] and [54] stated that MTC traffic types significantly depend on the type of technologies used for communication between devices and MTC application domains. This subsection presents MTC service requirements by focusing on 3GPP networks.

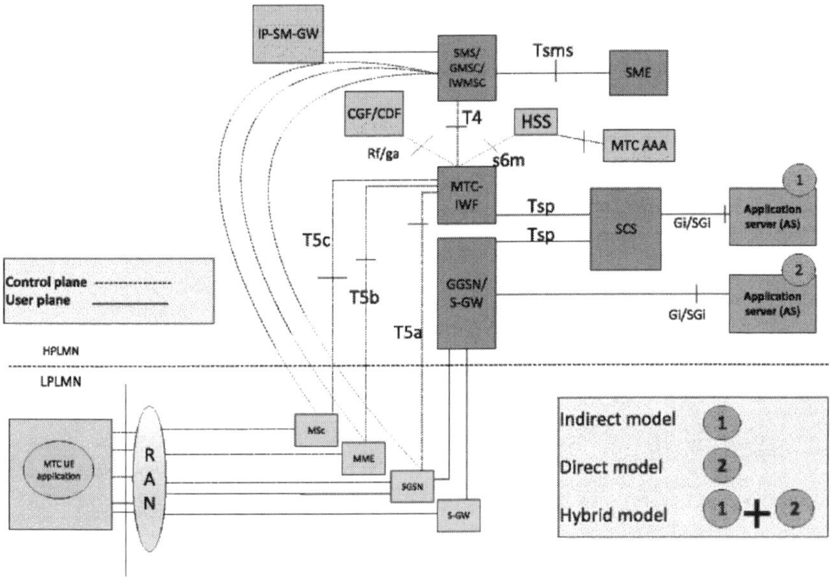

Figure 12 3GPP defined E2E network reference model for MTC (re-drawn from [143])

3GPP MTC features The MTC traffic characteristics differ significantly from application to application. Resultantly, system as well as service optimizations also vary, thus a particular optimization approach might not be applicable to the vast M2M application range [55]. 3GPP has defined a set of features for MTC in order to provide a structure to optimize systems according to varying MTC applications. In general, 3GPP identifies the following MTC features in its technical specifications TS 22.368 [55]:

1 Small-sized data transmissions

2 Infrequent mobile and data termination

3 Device low-to-high mobility

4 Time-controlled traffic behavior

5 Group-based policing and addressing

6 Ensuring secure connections and transmissions

7 MTC device monitoring requirements

Table 4 List of the reference points and network elements of the 3GPP reference model for MTC communications (taken from [45, 51–53]). The acronyms of network elements are given in Table 1

	Descriptions
Reference points	
Tsms	It enables the communication between an entity lying outside the 3GPP network and MTC using SMS
Tsp	It connects MTC-IWF to SCSs to provide several functionalities such as device triggering
T4	It performs SME tasks, e.g., connects MTC-IWF to the HPLMN domain. T4 also supports device triggering functionlities
T5a	It enables communication between MTC-IWF and serving SGSN
T5b	It establishes the communication path between MTC-IWF and serving MME
T5c	It creates the communication link between MTC-IWF and serving MSC
S6m	It performs subscription and MTC-related information tasks by establishing the communication path between MTC-IWF to HSS/HLR
S6n	It also performs subscription and MTC-related information tasks by establishing communication link between MTC-AAA and HSS/HLR and supports interrogation of HSS/HLR
Network elements	
MTC-IWF	One of the instances of MTC-IWF exists in HPLMN to support indirect and hybrid models
	Receives trigger request of an MTC device from SCS
	Reporting and updating SCS about the acceptance or rejection of the device trigger request
HSS/HLR	It determines whether an SCS is allowed to send a trigger to a particular MTC
GGSN/P-GW	It enquires an MTC AAA server for regenerating external identifier(s)
SGSN/MME/MSC	Accepting an MTC device trigger from MTC-IWF
	It confines the device trigger information in NAS messages sent to MTC
	It accepts a device trigger acknowledgement from the triggering MTC
	It reports to the MTC-IWF about the device trigger delivery success/failure status
	It also delivers SGSN/MME congestion/load information to MTC-IWF
SMS-SC	It delivers the routing information from MTC-IWF to SMS-GMSC
MTC AAA	It uses standard procedures such as RADIUS to identify the destination external AAA server
SCEF	It performs several tasks such as device authentication and profile management
	It makes external entities capable to discover various service capabilities

3GPP MTC common service requirements 3GPP has set certain service requirements for MTC in its Release 10 and onward. The major 3GPP-based MTC service requirements are summarized below (taken from [55–58]):

1 The system shall support a large number of MTC devices, i.e., connectivity for massive number of MTC devices.

2 The system shall provide an efficient resource allocation scheme to support MTC devices which send or receive small sized data, i.e., with longer inter-arrival/inter-send time.

3 The system shall provide an option for an MTC user for activating or deactivating MTC features.

4 The networks shall enable service providers to identify which MTC features are activated for a particular MTC user.

5 In the case of overload, the network shall restrict downlink transmissions and signaling.

6 In the case when a large number of devices attempt to access the network, the system shall support a scheme to limit peaks in the data as well as in signaling traffic.

7 If the network is overloaded, the system shall be capable to limit access towards a particular access point name (APN).

8 Each MTC device shall be capable to support the extended access barring (EAB) scheme defined in 3GPP TS 22.011 [27].

9 The system shall support methods to ensure low power consumption of an MTC device in order to increase its lifetime.

10 If MTC devices are not sending or receiving data, then the devices can or cannot be kept attached to the networks, depending on operator policies and requirements of an MTC application.

MTC device triggering 3GPP-based device triggering requirements of an MTC device are summarized as follows [55, 58]:

- Whenever an MTC server shows a trigger, an MTC device shall initiate communication with an MTC server.
- The network shall ensure that an MTC server's triggering indications to trigger MTC devices is authentic.
- In case of triggering from a source which is not authentic, the network shall be able to report to the MTC device by providing the details of fictitious/malicious source.
- The network should provide a mechanism to give information to MTC devices regarding authentic servers or sources.
- The MTC device shall create a link to communicate with an MTC server. Therefore, an MTC device shall receive a trigger whenever it is indicated by an MTC server. There can be three possibilities, i.e., receiving trigger indications while an MTC device is not attached to the network, receiving trigger indications when an MTC device is attached to the network without transmitting/receiving data, and receiving trigger indications when the MTC device is attached to the network with an established data connection.

MTC addressing and identifiers [55, 58] Each MTC device shall be assigned a private IPv4 address by the mobile network operator (MNO) if the MTC server exists in a public IPv4 address space. Regarding identifiers, the system shall uniquely recognize the attached MTC users. Furthermore, the system shall present techniques for the network operator to efficiently manage MTC-related numbers as well as identifiers.

1.4.4 Standardization efforts for M2M communications

According to [15], M2M communications significantly rely on numerous technologies spanning across several industries. Due to the wide application range, standardization in M2M communication is more challenging than the traditional standardization for a particular platform. Consequently, the required standardization scope is much higher than the existing standardization developments and entails to upgrade/modify existing standards to support future M2M traffic on a large scale. 3GPP, ETSI, IEEE, and Telecommunications Industry Association (TIA) have played a significant role in developing new standards to support M2M communications. The standardization work related to M2M communication is on-going in several standardization organizations outside 3GPP (e.g., ETSI, IEEE, TIA, and oneM2M). Moreover, these standardization bodies have also described the required enhancements in the existing communication standards in their specifications. One important effort for M2M standardization in 2012 is the *oneM2M* partnership by the global standard development organizations (SDOs) [59].

This section presents the major standardization efforts made by several SDOs to facilitate M2M communications and services. The major efforts made by several standardization bodies are described below. Moreover, a summary of their major contributions is given in Table 5.

oneM2M [60] oneM2M is a global initiative to support future M2M communications. Eight of the world's primary ICT standardization organizations such as ETSI and TIA have initiated oneM2M. Under this oneM2M umbrella, six of the globally active industry consortia such as broadband forum, OMA, and Continua have joined the regional standards bodies for ensuring efficient M2M communications. Other than the standards bodies and industry consortia, around 200 members from universities, research labs, and big companies have also joined this initiative. One of the primary objectives of oneM2M is to develop a technical design which should fulfill the requirements for a general service layer platform for M2M communications. Consequently, the designed specifications can be easily implanted within several hardware- and software-based systems to offer connectivity for a massive number of M2M devices to the backend server worldwide. Additionally, oneM2M focuses on to influence other organizations offering numerous M2M services in the perspective of business such as ITS, logistics, e-healthcare/m-healthcare, industrial automation, and smart metering and monitoring.

Table 5. Summary of contributions made by the major standards development organizations to support M2M communications

SDOs	Contributions	Reference
oneM2M	• M2M functional architecture, requirements, and security solutions	[146, 147]
	• Service layer core protocol specification	[148]
	• Protocol binding (CoAP, HTPP, and MQTT)	[149–151]
3GPP	• MTC reference model	[51]
	• MTC service requirements	[51]
	• Radio resource management for MTC	[10, 61]
	• Considering MTC security aspects	[62, 63]
	• Describing MTC-use cases	[61]
ETSI	• M2M networks architecture and technologies	[28, 49, 65]
	• QoS for M2M communications	[67]
	• M2M service requirements	[67]
	• M2M-use cases: e-healthcare, smart metering, automotive, and smart grid	[69–72]
	• Several M2M-related interfaces, i.e., *mla,dla*, and *mld*	[66]
	• M2M data semantic support	[152]
	• SmartM2M test specifications: interoperability for ETSI M2M primitives	[68]
IEEE	• Optimizing air interface	[73, 74]
	• Use of sub-GHz spectrum for M2M communications	[74]
	• Optimizing air interfaces for the smart grid IEEE 802.15.4	[74]
GSMA	• M2M resource management schemes	[77]
	• M2M-related protocols and interfaces	[77]
	• M2M communications using remote SIM	[77]
CCSA	• Supporting smart grid applications	[81]
WFA	• Integrating Wi-Fi in smart grids	[79]
	• e-healthcare application scenarios	[79]
WiMAX	• M2M deployment scenarios	[78]
	• Functional and network improvements	[78]
OMA	• M2M device management according to ETSI remote entity management	[80]
	• Personal area networks for ensuring efficient capillary M2M communications	[80]

3GPP 3GPP has made progress for network developments for inter-MTC device communication in Release 12. 3GPP proposed several solutions to handle large numbers of MTC devices in GSM and UMTS networks. The study focused on handling a massive number of MTC devices which are expected to communicate either directly with each other or with the backend server simultaneously. Moreover, the service requirements for MTC are addressed in [54]. Several key issues such as network congestion due to the large number of devices are discussed. In addition, numerous possible solutions such as group-based communication are given in order to tackle the aforementioned issues. 3GPP in its Release 12 [61] described various enhancements as well as use cases for MTC. Several applications for MTC such as transportation and logistics, surveillance, and e-healthcare are addressed. The security aspects in MTC are addressed in [62] and [63]. Furthermore, a detailed study of RAN enhancements and GERAN improvements for MTC is also given.

In addition to the aforementioned contributions, 3GPP started interacting with third parties in [64] and empowered them to facilitate users through third party services using 3GPP functions. In order to accomplish it, third parties also enable MTC services after making an agreement with operators. Furthermore, 3GPP has started working on an MTC architecture in Release 10. In Release 11,

3GPP defined several interfaces between the core network (CN), MTC service, and application enablement platforms. 3GPP is currently working on small data transmissions and low power consumption devices in Release 12 SA2. The major existing and expected efforts to support MTC in upcoming Releases are summarized in Table 6.

Table 6 Overview of 3GPP MTC standardization efforts (based on [145])

Parameters	Work items
3GPP Release 11 (2012)	• Power preference indications • Random access control overload
3GPP Release 12 (2014)	• Cat-0 MTC devices (low cost) • Power-saving modes • Device assisting information for eNB tuning
3GPP Release 13 (expected in 2016)	• Low-cost MTC modules • Coverage enhancements • Power-saving enhancements for eNB tuning

ETSI ETSI has made serious efforts in developing standards and protocols for M2M communication. ETSI has established a new committee with goals to provide an overall M2M communication overview. Moreover, the group has also focused on relating M2M communication closely with the ETSI development for the latest communication standards. Moreover, ETSI focuses on to relate M2M communication with the advanced 3GPP new communication technologies. The detailed E2E network architecture for M2M communication by ETSI is given in [28]. Furthermore, a detailed study of the inter-communication of the proposed ETSI M2M architecture with local area networks such as ZigBee is given in [65]. In addition, several communication interfaces such as *mla*, *dla*, and *mld* for M2M communications are provided in [66]. Moreover, the future requirements for M2M communication such as mobility support are discussed in [67]. Test specifications for the constraint application protocol (CoAP) binding of M2M primitives is presented in [68]. In addition, several use cases of M2M communication such as e-healthcare and smart metering are explained in [69] and [70], respectively. Moreover, M2M automotive applications and smart grid applications are discussed in [71] and [72], respectively.

IEEE IEEE has started studying M2M communication by announcing working groups for M2M communication [73, 74]. The working group *IEEE 802.16p*

10/0005 describes the usage of M2M communication and various standard modifications which are required to support future M2M communications. In addition, the working group *IEEE 802.16p-10/0014* provides various methodologies which seem to be helpful for M2M communications. Furthermore, IEEE also focuses on air interface optimization for mass device communication, the use of sub-GHz spectrum (Wi-Fi) for M2M communication, and optimizing air interfaces for the smart grid IEEE 802.15.4.

TIA [75] This SDO provides bi-directional communication between smart devices and the networks in its TR-50 framework [76]. Furthermore, it provides an extension for e-healthcare scenarios in which e-healthcare sensors can access the converged personal network service (CPNS) server in the global network via a mobile gateway. This extension in e-healthcare provides monitoring and simple daily care services. Hospitals and pharmacies can access information received from the sensors deployed on the body of the patient.

GSM Association (GSMA) [77] This SDO focuses on GSM-based M2M communication in which the major developments include module design, radio interface, and authentications.

Worldwide Interoperability for Microwave Access (WiMAX) [78] This SDO covers developments mainly in deployment models, functional improvements, and network improvements.

Wi-Fi Alliance (WFA) [79] WFA covers integration of Wi-Fi within smart grids and e-healthcare cases.

Open Mobile Alliance (OMA) [80] The key enablers for M2M devices and services include M2M device management, M2M gateway management, personal area network services to fulfill ETSI M2M area network requirements, and location detection services [80].

China Communications Standards Association (CCSA)[81] CCSA focuses on standards for interfacing smart grids and location-based services.

1.4.5 Major research projects

Due to the emerging use of mobile networks for M2M communications, several research projects such as EXpAnding LTE for Devices (EXALTED) and Achieving LOw-LAtency in Wireless Communications (LOLA) focused M2M communications for supporting increasing M2M traffic efficiently. Furthermore, several of the latest EU research projects such as METIS-2020 have extensively addressed existing and upcoming M2M-related issues in their research. Consequently, several recommendations and solutions have been proposed in order to handle M2M small-sized transmissions efficiently on a large scale. METIS-2020, one of the largest EU research projects addressed M2M upcoming issues as a front line research topic, within the list of its horizontal topics.

Resultantly, several solutions have been proposed for handling future mobile M2M traffic efficiently. This subsection presents an overview of these projects and highlights their major contributions, for the ease of the reader to understand the existing and upcoming issues and research gaps. In the end, a summary of all major contributions related to M2M communications is given in Table 7.

Table 7. Summarizing contributions of the major research projects for M2M communications

Project name	Duration	Contributions	Reference
METIS 2020	Nov. 2013–May 2015	• Scenarios, requirements, and KPIs for future M2M	[86]
		• Requirements for new air interfaces	[83]
		• Novel radio link concepts for M2M	[153]
		• Radio resource allocation techniques	[153]
		• M2M traffic offloading ideas/low signaling overhead	[154]
		• M2M via D2D communications	[153]
		• M2M massive random access	[154]
EXALTED	Sept. 2010–Mar. 2013	• Impact of M2M applications on business models	[155]
		• LTE-M algorithms and procedures	[156, 157]
		• EXALTED M2M system architecture	[158]
		• Optimizing a Linux operating system for M2M	[159]
		• Device management	[160]
		• Capillary M2M traffic aggregation	[161]
		• Security solutions for M2M	[162]
		• Security solutions for P2P relaying	[163]
LOLA	Oct. 2010–Mar. 2013	• Evaluation of M2M application scenarios	[164]
		• Access layer techniques for low latency robust transmissions	[165]
		• Spectral efficiency with M2M	[166]
		• M2M in long-range and short-range mesh networks	[165]
		• Transmission technologies for M2M gaming services	[165, 167]
		• PHY/MAC design procedures, algorithms, and scheduling	[165, 167]
		• Characterization and modeling of traffic	[167]
		• Experimentation traffic measurement testbench etc.	[168]

METIS-2020 [82] Mobile and Wireless Communications Enablers for the Twenty-Twenty Information Society (METIS-2020) is an EU research project, initiated in 2013. Twenty six partners including major operators such as Ericsson, NTT Docomo, and NOKIA have participated in this project. The primary goal of this project is to lay the foundation of upcoming 5G mobile and wireless systems. Since cellular-based M2M communications is one of the latest research topics, several massive machine communication (MMC)-related issues are addressed in METIS [83]. Furthermore, future MMC scenarios are discussed, and consequently several novel radio link schemes are proposed to handle the increasing mobile M2M traffic. Since handling MMC in future mobile networks is a challenging task especially for the radio link layer, METIS focused on providing multiple access and multiple control schemes, advanced MAC and HARQ algorithms, synchronization issues, and MMC-related signaling limitations for MMC scenarios in its radio link analysis.

EXALTED [3] EXALTED is an EU FP7[4] ICT research project commenced in 2010 (September) for a duration of 30 months. The project was coordinated by Sagemcom SAS, France, along with the other 12 partners including well-known operators and module manufacturers such as Vodafone, Ericsson, and Gemalto. One of the key issues of EXALTED is to provide secure, cost- and energy-efficient M2M communications in the context of future mobile technologies. Therefore, EXALTED also considered existing challenges in mobile networks to finally deliver the foundation of a new scalable network architecture. Furthermore, the project aimed at the specifications of massive machine communications under the name of *LTE-M* networks particularly in mesh networks. The resulting system provides a low cost and spectrally efficient M2M system with extended coverage in LTE networks. Approximately 1000 devices were expected to be supported per cell with the proposed architecture and schemes by assuming quite relaxed latency requirements greater than 100 ms.

LOLA [84] LOLA is another major EU FP7 research project coordinated by EURECOM, France. The project focused on access layer enhancements to support M2M communications in mesh networks with medium coverage as well as in mobile networks such as LTE-A with long range. LOLA aimed to ensure efficient spectrum utilization by targeting a particular set of applications with low latency requirements such as gaming. In cellular networks, the project focused on improving transmission technologies for supporting gaming applications which is expected to be one of the crucial areas for operators to increase their revenue in the future. Therefore, the M2M-related research focused to deliver recommendations to 3GPP regarding PHY/MAC functionalities in support of M2M. LOLA provided inputs to 3GPP standardizations directly as well as targeting the improvement of LTE MAC and PHY layers by designing schedulers, framing, adaptive modulation/coding schemes, as well as ACK/NACK feedback.

1.5 Mobile MTC challenges and solutions

Existing mobile networks such as LTE were originally designed and subsequently upgraded due to the tremendous growth in mobile data traffic. On the one hand, the CISCO 2014–2019 forecast [85] shows that mobile data traffic will grow to 24.3 exabytes (EB) per month by 2019, up from 2.5 EB in 2014. On the other hand, M2M traffic is expected to increase even more rapidly as compared to traditional mobile traffic, making QoS provisioning a huge challenge. Consequently, current mobile networks will not be able to support the increasing number of M2M devices worldwide [83]. M2M traffic poses serious challenges to cellular operators due to the high density of devices transmitting small-sized payloads [86]. For instance, according to the latest M2M traffic forecasts, GSMA 2020 forecasts that the number of cellular-based M2M devices will grow to one billion by 2020 [87]. Furthermore, approximately 3.2 billion cellular-based M2M devices are expected in 2024 (M2M Magazine 2014–2024 forecast [88]). This rapidly increasing M2M devices would undoubtedly cause severe challenges for mobile networks. The major challenges for mobile networks include the following:

- Support for a large M2M device volume
- Small-sized data transmission after regular as well as irregular time intervals
- Ensuring low power consumption as compared to non-cellular standards
- Low latency, high reliability, and support for various mobility profiles
- Ensuring that M2M traffic is not affecting regular H2H/H2M traffic

The remaining part of this section discusses several existing and upcoming issues and challenges posed by M2M communications in LTE/LTE-A networks.

1.5.1 LTE-A inefficient radio resource utilization

Existing mobile standards are neither designed to handle small-sized payloads efficiently nor to support large number of devices simultaneously [83]. Consequently, this will lead to network congestion. Current mobile networks must be extended or revised to support massive M2M devices in the future. Since mobile radio resources are valuable assets and scarcely available, it is required to ensure efficient utilization of radio resources for M2M communications. According to present 3GPP standardization, the smallest unit of the radio spectrum allocatable to a single device is 1 PRB which is capable to transmit several hundred bits under favorable channel conditions (see TB size for 1 PRB in Fig. 7). However, allocating 1 PRB to a single M2M device could significantly degrade radio spectrum utilization. This is due to the fact that the capacity of a PRB can be much higher than the actual size of the device payload.

Figure 13 illustrates how inefficiently a PRB can be used when a device sends small-sized packets experiencing good channel conditions. Two exemplary device payloads (i.e., 4 and 8 bytes) are considered. Moreover, a PRB utilization efficiency is evaluated in terms of percentage of the actual payload and the padded zeros with respect to the MCS index. It can be seen from Fig. 13 (left) that with the lower MCS values, e.g., 0–6 depicting unfavorable channel conditions, the PRB capacity is smaller, and therefore it is fully utilized to transmit the given payload. On the other hand, with the higher MCS values, e.g., 20–26 representing favorable channel conditions, the capacity of the PRB increases. However, small packets with increasing percentage of padded zeros are transmitted. Similarly, Fig. 13 (right) depicts the percentage of PRB utilization and padded zeros for the second scenario. However, it can be seen from Fig. 13 (right) that for the lower MCS values, the PRB is used efficiently as compared to Fig. 13 (left). However, with the higher values of MCS, the percentage of padded zeros is still significantly large. This shows that approximately 10–20 times more packets can be accommodated per PRB under favorable channel conditions in the given exemplary scenarios.

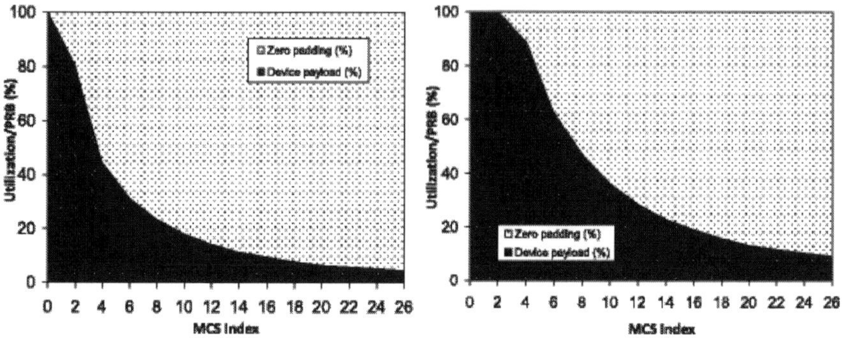

Figure 13. An illustration of inefficient PRB utilization for M2M communication with device payloads of 4 bytes (left) and 8 bytes (right)

Due to the inefficient utilization of scarce radio resources, M2M traffic is expected to degrade the performance of traditional mobile traffic. For instance, the authors of [89, 90] evaluated the influence of increasing M2M traffic on LTE data traffic through system level simulations using an LTE-A-based model developed in the OPNET Modeler. The authors concluded that the performance of LTE/LTE-A low priority traffic such as FTP and HTTP is degraded significantly when the number of M2M devices was raised to 750 (Fig. 4 in [89]). This happened due to the fact that e-healthcare traffic was prioritized over LTE traffic.

In LTE/LTE-A networks, radio resources are distributed among several users in order to accomplish user plane and control plane services. The massive number of M2M devices is expected to influence the performance of LTE/LTE-A networks for both the user and the control plane due to the exchange of excessive signaling messages for sending/receiving small packets. Therefore, these topics demand serious attention to enhance M2M services efficiently using LTE/LTE-A networks.

1.5.2 MTC radio resource allocation mechanisms

In communication networks, efficient radio resource management has always remained a challenging topic. It has become more challenging for mobile networks to handle traditional traffic co-existing with cellular-based MTC traffic. Therefore, new and optimized radio resource management schemes are required to support future MTC traffic, for which LTE and LTE-A networks are not yet optimized [83]. Several proposals have been presented so far for efficient radio resource allocation for LTE/LTE-A networks.

For instance, authors of [18] proposed two methods for allocating radio resources between MTC and H2H devices. The first method is based on the orthogonal radio resource allocation. The strength of this resource allocation scheme is its simplicity, whereas the drawback is the low spectral efficiency from the system level aspect. However, the second method, i.e., shared radio resource allocation improves spectral efficiency of the network in a way that MTC devices can reuse the bandwidth allocated to H2H/H2M users. However, the drawback is an increased level of interference as compared to the former approach. In 3GPP LTE/LTE-A networks, radio resources are split into PRBs also called subchannels in the context of resource allocation. For instance, if MTC traffic co-exists with traditional H2H traffic in a communication environment, five distinct links are available which are listed below, see Fig. 14 (taken from [18]):

- eNB to UE (eNB → UE) link
- eNB to MTCD (eNB → MTCD) link
- eNB to MTCG (eNB → MTCG) link
- MTCG to MTCD (MTCG → MTCD) link
- MTCD to MTCD (MTCD → MTCD) link

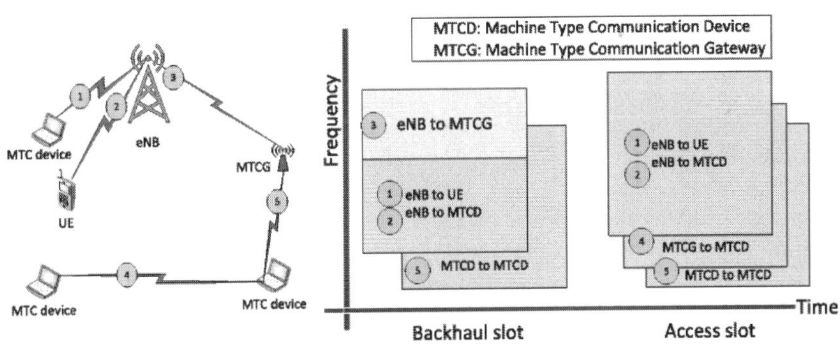

Figure 14. Radio resource separation in LTE-A networks with MTC. Transmission links for MTC and regular mobile traffic (*left*) and an example of radio resource partitioning (*right*). Re-drawn from [18]

In a mixed traffic environment, one of the major challenges is handling of interference. Therefore, it requires a sophisticated resource partitioning mechanism. The radio resource partitioning enables radio resource allocation among different devices in a coordinated way. The restrictions can be either on the basis of available PRBs or transmit power. As a result, such restrictions provide the possibility for improving the SINR ratio.

Several scheduling algorithms have been proposed for LTE/LTE-A networks. For instance, authors in [91] evaluated the performance of dynamic packet scheduler for LTE networks. The advantage of this approach is the effective control of throughput and fairness between users. However, the drawback of the proposed scheme is that it does not differentiate services. Additionally, authors in [92] introduced a principle of metric decoupling to maximize throughput. Moreover, authors concluded that the fairness among users can be controlled efficiently either through weighting metric in the frequency domain or handling priority in the time domain. In [93], a path selection criteria for small cells is proposed for joint transmissions on the access and backhaul links in the millimeter wave band to improve the performance of device-to-device (D2D) communication. The proposed D2DMAC scheduling scheme is centralized and outperforms traditional approaches particularly in terms of delay and system throughput. Authors in [94] studied dynamic TDD transmission mechanisms for small cells with several interference management schemes such as interference cancellation, cell clustering, as well as power control. The advantages of proposed schemes are potentially large in low-to-medium traffic loads. Boosting UL power in dynamic TDD scheme significantly outperforms static TDD scheme. Approximately 30–60 % and 210–300 % user throughput is improved in DL and UL, respectively. Moreover, authors in [95] proposed a feasible strategy to enable dynamic TDD transmissions in small cells to achieve UL and DL throughput gains. Additionally, authors in [96] made a theoretical analysis on the SINR performance in case of dynamic time-division duplex (TDD) transmission in small cells. By introducing a partial interference cancellation scheme, outage probabilities are notably decreased. Moreover, authors concluded that the risks of link failures can be significantly reduced by canceling one interfering cell in case of low-to-medium loaded scenarios.

Authors in [97] proposed a cell breathing framework in downlink to minimize breathing effect in femtocells, and thus to balance data rates and cell coverage. A voting-based mechanism, i.e., femtocell virtual election rule (FEVER) is proposed in which users notify the femtocell base station about their current channel conditions which significantly improve system efficiency and fairness among users. In [98], authors studied game dynamics and learning schemes and proposed a novel learning scheme called *cost-to-learn* which integrates factors such as switching cost and delay as well as user behavior. Resultantly, users concurrently learn their optimal payoff and strategy with the designed CODIPAS-RL (combined fully distributed payoff and strategy reinforcement learning) scheme. Moreover, authors created *asymptotic pseudo-trajectories* to solve differential equations. In [99], authors proposed a distributed and coordinated scheduling scheme for LTE/LTE-A networks where each cell tends to allocate MCS, PRB, and transmit power to users for minimizing DL power while satisfying user throughput demands. Furthermore, each cell notifies neighboring cells about PRB allocation to cell-edge users. Resultantly, the proposed scheme provides stable solutions for interference mitigation as well as efficient frequency reuse. Moreover, a simple self-organizing scheme is proposed in [100] in which cell transmit power is minimized in order to utilize frequency reuse patterns efficiently in distributed cellular networks.

To support MTC traffic, authors proposed in [101] various scheduling schemes by considering channel conditions as well as delay constraints, thus to maximize the number of supported devices per cell. In [21, 102], authors proposed clustering techniques for MTC traffic. MTC devices are classified into clusters with similar QoS requirements. Fixed AGTI is allocated to each cluster depending on its priority as well as traffic flow. The advantage of the proposed scheme is the capability to support the maximum number of devices while considering their priorities and delay restrictions. The performance of the above mentioned proposals are rather promising at the expense of high signaling overhead [103]. Additionally, there are several limitations in the aforementioned proposals such as the constant MTC traffic flows were examined (contrary to potentially random traffic). In addition, the mean delay experienced by end users is expected as a performance indicator instead of the probabilistic packet delay and loss rate which are recognized as more realistic performance indicators. The authors of [103] attempted to overcome these issues by optimizing the signaling overhead as well as by considering random MTC traffic. The signaling overhead is reduced by merely considering MTC QoS class identifiers (QCIs) which are known in the beginning of a connection and by avoiding feedback messages regarding the buffer status and channel conditions. However, considering various MTC traffic growth forecasts, efficient radio resource management is still one of the most challenging research tasks and demands serious attention [104]. New and advanced schemes are required in order to allocate radio resources efficiently to handle massive MTC traffic in the future.

1.5.3 M2M random access challenges in LTE-A

In LTE/LTE-A networks, MTC devices contend for PRBs in order to transmit and receive data using an RA procedure. The wireless access schemes can be based either on capillary short-range networks such as WLAN, ZigBee, and IEEE 802.15.4x or long-range mobile networks such as GPRS, GSM, UMTS, and LTE/LTE-A. The wired access techniques are more reliable, introduce small delay, and provide higher throughput. However, it is not suitable for several MTC applications due to various factors such as lack of scalability, cost efficiency, and mobility. In order to cope with these issues, wireless networks play a prominent role. Capillary wireless access is usually used for short-range links and is less expensive, scalable, and requires low power. However, several factors such as low data rates, weak security, higher interference, and less coverage make it less suitable in most of the MTC applications. In order to provide wider coverage, higher data rates, mobility support, roaming services, and security, mobile networks are more attractive and applicable for future MTC [19]. LTE-A is the latest mobile standard which is a potential option to support MTC. Since MTC devices access the network simultaneously and contend for

PRBs, this paper focuses on the contention-based RA challenges and possible proposed solutions.

PRACH overload control Per cell, a massive number of MTC devices is expected to access the network simultaneously [19]. Consequently, it causes a low RA success rate and a high PRACH network congestion. Moreover, unexpected delay, increased packet loss, high power consumption, and inefficient utilization of radio resources will occur. In addition, the PRACH will be further loaded when the increased number of devices will attempt to re-access the network after collisions. Therefore, efficient overload control mechanisms are needed to handle future MTC traffic efficiently.

QoS provisioning In a mixed communication environment, MTC and H2H traffic compete for PRACH resources. However, MTC traffic should fulfill its QoS requirements without degrading the performance of traditional H2H traffic [19]. Since MTC devices are capable to access the eNB either directly or using an intermediate gateway, the decision of selecting an eNB or a gateway can be taken based on the signal strength or channel gain particularly in the case of RA (discussed in Section 1.4.2). In addition, this decision should consider QoS performance based on the contention-based procedure. For instance, a device can receive satisfactory signal strength through direct communications with an eNB. However, if the device uses an intermediate gateway, it can potentially increase RA success probability, reduced power consumption, as well as smaller delay. Furthermore, optimum selection of transmission mode should also satisfy MTC QoS requirements such as lower delays, e.g., in the case of emergency/accidental information, and higher data rates, e.g., in surveillance applications. Therefore, the selection of the transmission mode is also crucial. Unfortunately, most of the existing literature does not consider the point selection mechanism for overload control in the RA procedure.

Cognitive M2M communications One of the major issues is the increasing signaling overhead due to the huge number of MTC devices per cell. This increases the possibility of signaling congestion in the network [105]. In addition, there may not be enough bandwidth in order to support and satisfy all devices simultaneously. One of the traditional approaches is to increase the number of eNBs. Furthermore, eNBs supporting MTC and non-MTC traffic should cooperate in order to reduce the risk of network congestion. However, interference between MTC and non-MTC traffic would be a challenging task. Centralized coordination would be effective in order to limit interference. However, it increases the signaling overhead as well as the management complexity. Distributed resource management could be an effective approach in order to limit interference between MTC and non-MTC traffic. Thus, cognitive MTC communication is a promising idea which can be helpful in order to reduce interference between MTC and non-MTC traffic [106]. In addition, the opportunistic random channel mechanisms would be beneficial in order to avoid network congestion. Therefore, enabling cognitive radio capability within an MTC device might be a promising research topic. To handle a large number of devices per cell, random access mechanisms would be beneficial in a way that

MTC and non-MTC devices would share the same limited radio spectrum, thus compete for radio resources.

Group-based M2M communication Due to the increasing MTC devices, major challenges are the allocation of RA slots and signaling overhead. In such cases, group-based communication would play a vital role in managing increasing MTC traffic [21]. The primary objective of group-based communication is to reduce the signaling congestion on the air interface. A group head can be selected which collects uplink/downlink data packets, status information, etc. and then forwards the collected information to the corresponding eNB/MTC devices. Since efficient preamble allocation mechanisms are required to avoid collisions, game theory can additionally be a useful tool to optimize preamble allocation [107]. In addition to reduce the network congestion risks, it can also reduce the power consumption of MTC devices. This is due to the fact that machines need to send/receive data to/from a nearby point, e.g., a relay node. In addition, categorizing MTC devices into groups demands special attention as the grouping should be based either on logical or physical characteristics on the basis of, e.g., QoS, jitter, and other MTC traffic characteristics.

1.5.4 Random access overload control mechanisms

In order to control overload in LTE-A networks due to massive MTC traffic, the following solutions have been proposed for PRACHs:

Access class barring (ACB) scheme In a traditional ACB procedure, an eNB broadcasts access probability (AP) and access class (AC) barring time within a cell [19]. When a device initiates the RA mechanism, it draws a number between 0 and 1. The result of the draw is compared with AP. If the result is less than AP, the device can proceed with the RA mechanism. Otherwise, device has to wait for a duration of AC barring. In Release 10 and onward, 3GPP has proposed an extended access barring (EAB) mechanism in which a certain class of delay tolerant traffic cannot perform the RA procedure whenever an EAB is initiated [108]. The eNB can control the PRACH overload by decreasing the AP value. Resultantly, it lowers the RA attempts. However, it might increase the RA delay for certain devices. Furthermore, in a cooperative ACB mechanism [109], an eNB selects the value of ACB depending on the level of network congestion. An optimization problem is considered to balance the number of MTC devices attached to a single eNB. However, the drawback of the proposed approach is that it does not consider the priority of MTC devices. In order to cope with this limitation, a priority-aware RA (PRA) scheme is proposed in [110] which takes into account the priority of MTC devices. PRACHs are pre-allocated to multiple MTC classes by adopting a class dependent backoff procedure.

Resource partitioning mechanism The resource partitioning mechanism illustrates the separation of orthogonal PRACH resources/preambles between H2H and MTC traffic [19]. Two resource partitioning mechanisms are proposed in [111]. In the first method, the preambles are split into two subsets, and each subset is dedicated to MTC and non-MTC devices. In the second method, preambles are again split into two subsets. However, one subset is particularly dedicated to non-MTC traffic, and the second subset is shared by MTC and non-MTC traffic. Once the ACB mechanism is passed successfully, the devices can send preambles using RA. However, the drawback of the proposed approach is that it neglects the eNB selection method and backoff procedure.

Slotted access scheme In this scheme, each MTC device receives dedicated access slots to perform the RA mechanism and, other than this, always remains in sleep mode [108]. Using ID and RA cycle number, each device can calculate its allowable access slots. An eNB broadcasts the RA cycle number which is a multiple of radio frames. The advantage of the given scheme is that MTC devices can share the access slot if the total number of unique access slots is less than the number of devices within a cell. However, the drawback of this scheme is the large RA request delays if the RA cycle length is increased. However, it can significantly reduce the collisions at the cost of large RA request delays. In the LTE-A RA procedure, the influence of an increasing number of transmission attempts on throughput and delay of the slotted ALOHA-based preamble contention is also studied in [112].

Clustering mechanism for MTC In this scheme, each device can be associated particularly with one or more groups. The radio resources are allocated based on the grouping principle, e.g., MTC devices are grouped based on their QoS requirements [21]. Furthermore, grouping can also be performed based on applications or geographical location. In each group, a group head is selected which communicates with an eNB by relaying requests of the group members to an eNB (i.e., peer-to-peer communication). One of the primary advantages of this scheme is to minimize power consumption for MTC devices [113]. Moreover, it significantly reduces the signaling overhead and ultimately minimizes the risk of network congestion.

Dynamic radio resource allocation In this scheme, an eNB allocates PRACH resources among MTC devices on the basis of PRACH overload and overall traffic load in the network [19]. In a PRACH subframe, a part of the subframe is used by the eNB for performing the PRACH mechanism in order to satisfy QoS and delay requirements, and therefore cannot be used for data transmission. Authors of [114] formulated this issue by reducing the number of PRACH subframes. The expected RA delay is kept within a given delay limit. In addition, a self-optimizing algorithm is proposed in [115] based on the 3GPP proposed slotted access scheme. In the proposed approach, an eNB can dynamically change RA slots for transmitting preambles by taking into account the actual traffic load. The advantage of the proposed scheme is that it minimizes the risk of network congestion.

MTC-specific backoff scheme In this scheme, the MTC and H2H RA attempts are delayed separately in a way that a smaller value of backoff time, e.g., 20 ms, is given to H2H traffic and a larger value, e.g., 960 ms, is assigned to MTC traffic [21]. Consequently, this scheme is effective in low overload situations. Whereas in high overload situations, i.e., when more devices perform the RA mechanism simultaneously, this scheme is unable to tackle the network congestion issues. Moreover, the drawback of this proposed scheme is that it does not takes into account the priority of MTC traffic when co-existing with H2H traffic.

Pull-based scheme This centralized scheme handles situations when more devices attempt to access the network using the RA procedure [108]. In this scheme, an MTC server communicates with an eNB and requests to page MTC devices. When an MTC device receives a paging message from the eNB, it will initiate the RA procedure. Based on the PRACH load and spectrum availability, an eNB is capable to control the required number of MTC devices to be paged. However, the drawback of this scheme is the use of extra control channels to page a massive number of devices.

1.5.5 QoS provisioning for M2M communications

QoS provisioning is one of the major requirements in telecommunication systems. In the perspective of service provisioning, ITU-T defines QoS as a set of all service-related tasks associated with the service provider when a particular service is offered to the end user [116]. Additionally, QoS parameters are mainly used to evaluate whether or not technologies and offered services fulfill the expectations of end users in terms of quality, reliability, and availability [117].

Unlike traditional mobile traffic, QoS class categorization for M2M communications has remained one of the most challenging tasks for operators due to the fact that M2M applications cover a wide variety of use cases. Since mobile networks are expected as a strong candidate to support future M2M communications, there is a strong requirement for defining unique QoS class categorizations for differentiating M2M traffic, e.g., based on priority and delay. Nevertheless, a particular QoS class categorization for M2M traffic has not yet been defined in the literature [117]. Authors of [118] proposed various M2M QoS classes based on the traffic characteristics. On the other hand, 3GPP and ETSI are putting efforts into allocating specific QCI values to each category in M2M communications. In this survey, several M2M characteristics and traffic models are summarized in Table 8 with a provisional QoS categorization depending on the application type, data packet size, required priority, expected delay, and mobility demands. For instance, it can be seen from Table 8 that M2M devices mainly transmit small packets with an expected inter-arrival time distribution ranging from milliseconds to hours. Furthermore, M2M traffic can be classified based on the type of traffic, e.g., real or non-real time traffic. Table 8 depicts various M2M traffic features which can be used to classify M2M traffic into various classes. For instance, emergency alerting, i.e., accidental and/or critical e-healthcare information, is a real-time application which is delay sensitive and therefore demands strict priority. On the other hand, traffic generated by POS terminals is also real time; however, it demands low priority compared to the emergency alerting. In addition, smart metering and monitoring applications generate regular traffic patterns, therefore it can be considered as low priority traffic, but with higher data rate requirements.

Table 8. Existing QoS class categorization for M2M communications [118, 142, 169]

Parameters	Class 0	Class 1	Class 2	Class 3	Class 4
Use cases	e-healthcare, emergency alerting	Mobile POS (point of sale)	Intelligent transportation, streaming,	Smart homes, security	Metering, monitoring
Traffic types	Random, real time	Random, real time	Random	Random	Regular
Priority	Strict, high	Low-1	Low-2	Low-3	Low-3
Data size in bytes	Small	Small	Small	Small/medium	Large
Inter-send time	Seconds	Seconds	Minute	Minute ↓ hours	Seconds ↓ hours
Mobility	Low	Low	Low ↓ high	No	Low
Delay budget	Milliseconds [approx. 125 ms]	Milliseconds	Milliseconds ↓ minute	Minute	Seconds ↓ hours

1.5.6 Low-cost low-power device requirements

In mobile M2M communications, the cost and power consumption of an M2M module are of significant importance to draw a comparison between cellular-based and non-cellular-based M2M communication services [6]. Unlike traditional mobile data traffic such as VoIP and multimedia services which demand high mobility and have no firm limitations for price and battery time of a device, M2M devices demand low price and low power consumption. 3GPP has considered this issue in its Release 12 and introduces a low-cost M2M device category called Cat-0 device. The average price of this newly introduced Cat-0 device is approximately 40–50 % of traditional LTE devices. The complexity of the upcoming MTC modem in 3GPP Release 13 is expected to be 25 % compared to a Cat-1 modem. Furthermore, this new category device called *Sub Cat-0* will not support Tx diversity and MIMO operations. Moreover, narrowband RF design supporting 1.4-MHz bandwidth, data rates up to 200 kb/s and transmit power of 20 dBm would sufficiently fulfill MTC application requirements, thus will further reduce device complexity. The key reductions agreed in 3GPP Release 12 and upcoming Release 13 are summarized in Table 9.

Table 9. Optimizations of low-cost and low-power M2M devices in 3GPP Release 12 and upcoming Release 13 [6, 170–172]

Parameters	Release 8 (cat-4)	Release 8 (cat-1)	Release 12 (cat-0)	Release 13
MTC device bandwidth (MHz)	20	20	20	1.4
Peak data rate downlink	150 Mbps	10 Mbps	1 Mbps	200 Kbps with a transport block size (TBS) of 1000 bits unicast traffic
Peak data rate uplink	50 Mbps	5 Mbps	1 Mbps	200 Kbps with a transport block size (TBS) of 1000 bits unicast traffic
Duplex mode	Full duplex (opt.)	Full duplex	Half duplex (opt.)	Half duplex (opt.)
Max. device Tx power (dBm)	23	23	23	20
Modem complexity (%) compared to Cat-1	125	100	50	25

In M2M communications, devices have strict battery power constraints peculiarly in battery-driven M2M devices [6]. Moreover, it is inconvenient to provide services in locations without a direct energy source such as water meters, sensors used in intelligent containers, on vehicles for on-board security, on the body of animals for wildlife management, massive number of metering devices deployed, e.g., in a tower. In such cases, it is expected that M2M devices should work for a longer time period (several years) which is indeed a challenging task for module manufacturers as well as service providers. Furthermore, M2M devices spend most of the time in the idle state due to comparatively longer inter-arrival times. The idle state is designed as a low power state in the LTE network in which devices usually sleep to save battery and wake up periodically to check any system information (SI) update or DL packet arrival by listening to the paging message broadcast by the LTE network [119]. In the existing LTE standards, the maximum allowed paging cycle is 2.54 s [120] which is not optimal for M2M communication due to the fact that the inter-arrival time between traffic sessions might be in the range of seconds to minutes. Therefore, consideration of longer paging cycles are under discussion in 3GPP [121, 122]. However, longer paging cycles can increase the delay of DL packets. Therefore, the range of the paging cycle should be selected carefully to keep a balance between delay and power saving efficiency. Figure 15 summarizes the power consumption in different states of a device.

In order to further improve power consumption for M2M devices, several techniques are being considered by 3GPP in upcoming Release 13. Since downlink traffic can be delay intolerant, it is not possible to use longer TAUs. Similarly, power consumption can significantly be improved through coverage extension. This is due to the fact that there is always a trade-off between power consumption and reachability. In addition, Release 13 may address functioning of optimized RRC by optimizing signaling overhead. For instance, a remarkable increase in battery life for two AA-size batteries has been observed from 13 to 111 months by increasing the DRX cycle from 2.56 s to 2 min. The basic idea to improve battery lifetime is to further reduce the activity of the idle mode over the required level for data transmissions/receptions. Thus, the sensors should not have a regular paging cycle which means that devices should only be in active mode if there is any data to transmit.

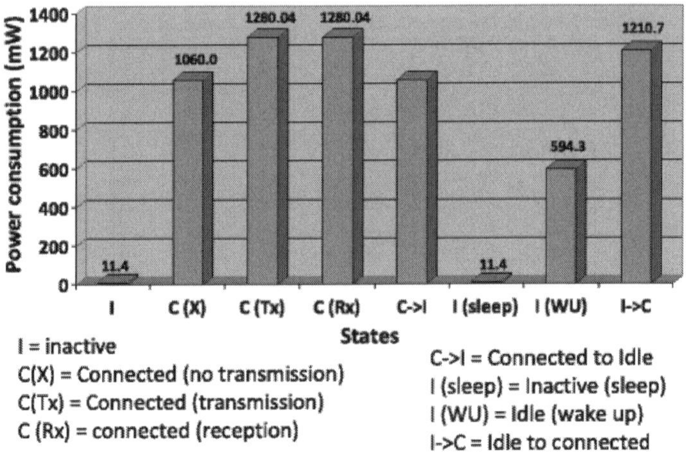

Figure 15 A comparison of power consumption for different device states [144]

1.6 M2M application areas

Mobile M2M communications cover a broad range of applications, e.g., e-healthcare, transportation and logistics, retail, safety and security, metering and monitoring, automation, and many more. This section discusses major M2M applications which can benefit from mobile networks such as LTE/LTE-A. Moreover, an overview of the major M2M applications covered in this survey is given in Fig. 16.

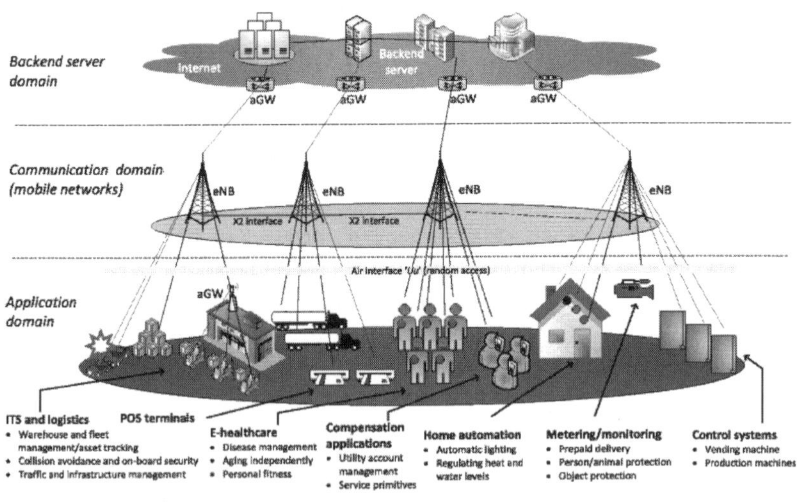

Figure 16 An overview of mobile M2M applications. Moreover, subcategories of major M2M applications are presented

1.6.1 Intelligent transportation systems and logistical processes

Due to the tremendous growth in the number of vehicles, transportation and logistics are considered as two of the main M2M communication domains worldwide [123]. ITS enables the exchange of information between vehicle infrastructure and ITS applications through several communication methods and technologies [5]. In ITS, vehicles communicate with other vehicles (vehicle-to-vehicle (V2V)) or with ITS infrastructure server (vehicle-to-infrastructure (V2I)) [50]. The major M2M applications in ITS include collision avoidance, safety and security, parking time management, Internet connectivity, and fuel consumption [3]. In addition, tracking and monitoring of vehicles are the main application areas of mobile M2M communication in transportation and logistics [5, 123]. Several research efforts investigated the support of M2M communications in ITS [5, 124]. Some of the major M2M applications in ITS are discussed as follows:

Collision avoidance and on-board security Collision avoidance and on-board security are the most significant M2M services in ITS. The main purpose of using mobile M2M communication is to deal with line-of-sight (LoS) limitations in order to avoid accidents [125]. In case of an emergency, the collected information is sent to other vehicles or infrastructure within the communication range. To avoid further incidents, communication between the server and the vehicles must be very fast to detect emergency messages and deliver warning messages immediately. Since the response time to emergency/warning messages is significantly small (i.e., in milliseconds), collision avoidance services demand high QoS and low latency. Furthermore, these warning messages are small in size, thus should only be sent in critical situations to ensure efficient utilization of scarce radio resources in mobile networks [126].

Traffic and infrastructure management All over the world, everyday drivers face the problem of road congestion that not only increases fuel consumption which leads to more emission and causes increased pollution but also causes stress for the drivers [127]. Traffic and infrastructure management can play a significant role in handling and controlling the problem of road congestion. ITS tackles the problem by providing bi-directional M2M communications, i.e., between vehicles and ITS server. Vehicles can send status updates about the position, speed, etc. to the infrastructure and can receive information/instructions about road accidents, emergency braking systems etc., from the infrastructure. Therefore, a better managed infrastructure improves the productivity and reduces the factors of cost and pollution in the society. Nevertheless, these applications do not demand high data rates due to the fact that only the abovementioned parameters are mainly sent [128].

The breakdown call (bCall) The bCall service simply starts a voice call and sends the current location/position of a vehicle to an adjacent roadside organization [5]. A bCall service is activated through a triggering operation which is done by the car by pushing triggers. Besides the bCall, an *enhanced* *bCall* service provides current diagnostic information along with the location of the vehicle. This allows for a remote diagnosis of faults and enables appropriate and in-time actions to be taken.

Additional M2M services in ITS In addition to the aforementioned applications, M2M communications in ITS support several operations such as tracking of a stolen vehicle, traffic reports and route planning, as well as infotainment services [5]. For instance, to recover a stolen vehicle, stolen vehicle tracking (SVT), service providers request data about the location from Telematic Control Unit (TCU) located inside the vehicle. In addition, drivers are also updated by sending reports regarding traffic in a particular region so that they can change/plan new routes in case of a traffic jam or an emergency. Furthermore, infotainment services aim to provide news/information to drivers and passengers through mobile TV, web-browsing, etc.

M2M in logistical processes Fleet management is one of the major applications of automotive M2M communications in logistical processes [123]. The movements of vehicles, containers, buses, cars, etc. are being tracked regularly through M2M devices which collect data, e.g., of the location, vehicle speed, temperature, distribution progress, fuel consumption, etc., and send this information to monitoring servers. Through regular monitoring, several activities of the system can be performed in an efficient way. For instance, the goods which are transported from one place to another are monitored regularly in order to accomplish in time delivery and to handle any undesirable situation during shipment processes. In summary, M2M communication addresses the following main issues in logistics, (1) *Global Connectivity*: the cargo moves across several regions/countries and thus must be monitored in order to stay updated; (2) *Reporting*: with the help of reporting, the exact location of the freight and the conditions of the objects can be easily monitored; and (3) *Alarming*: in applications such as warehouse management, fleet management, robotics, and control systems alarms are used to detect critical or emergency situations. Besides, M2M communications in logistics can provide additional services such as decreased operational cost, high inventory flexibility, increased supply chain visibility, and reduced loss of vehicles and containers [129].

1.6.2 e-healthcare

ITU in 2011 reported the increasing use of M2M services for medicine and electronic health records in e-healthcare [130]. The applications of M2M communications in e-healthcare include, e.g., remote monitoring of patients, provoking of alarms for handling serious conditions, and remote control of certain medical parameters. A patient is equipped with numerous M2M sensors which collect several health indicators of a patient, e.g., blood pressure, temperature, and heart rate. These devices transmit collected data to the monitoring centers such as hospitals, and clinics which store the recorded data and reacts accordingly [69].

Generally, the recorded data is mainly sent to medical centers by using WLAN technology, which is reliable for both the patients and medical control units. However, the lower coverage areas of WLANs may generate problems for the patients in case of mobility outside the home. If the patient is outside the home, e.g., for exercising, the regular monitoring of the patient is not possible due to the limited WLAN coverage. Consequently, in case of any unusual situation, the patient will not be able to get emergency aid by the medical center. In these situations, there is a need to extend the coverage and high mobility support features using mobile networks such as LTE-A. Mobility is one of the fundamental requirements of many M2M applications such as e-healthcare, ITS, and logistics. In this part, some of the major e-healthcare applications are discussed as follows [69]:

Disease management Remote monitoring and management of a disease are two of the common use cases of M2M communications in e-healthcare. Examples of disease management include diabetes management such as tracking the level of blood sugar and regulating the level of the insulin dosage. Furthermore, unusual heartbeats are recorded and sent to the medical center in order to monitor and analyze cardiac arrhythmias. Per time unit, the amount of recorded information as well as the frequency of reporting is mainly dependent on the type of disease to be managed. In addition, alarm functions are ensured in some disease management processes to tackle any undesirable situation by sending critical information to the doctor either on regular or on-demand basis.

Independent living with M2M M2M e-healthcare applications allow the people to live independently by monitoring numerous health parameters such as pulse rate, temperature, weight, blood pressure, and sugar level. This also reduces the number of required visits to a medical center. In addition, doctors/care center management can also track whether or not the patients are following the set schedule for taking their medications. Like disease management, sensors acquire data of various indications and then send this to the doctor in order to provide an interface to the patient to receive attention.

Fitness and better health M2M devices are usually deployed to collect/record health concerning parameters such as heartbeats, decreasing rate of fat level, respiratory rate, e.g., during exercise times and jogging. Furthermore, these M2M applications can be used to register several tasks such as duration and frequency of training and exercises. The physician can also make the health profile of a patient by recording data via M2M devices. Furthermore, a doctor/physician can also utilize this information to ensure delivery of precise and quick exercise programs, physiotherapy, etc. and can give comments/feedback to the patients about their recovery, further exercisel plans, etc.

1.6.3 Smart grid: M2M metering and monitoring services

Smart metering and monitoring are considered to be one of the main driving forces for increased M2M traffic [131]. Moreover, M2M communications play a vital role for improving the performance of the smart grid which is used to improve the performance of energy generation, distribution, as well as consumption. Smart metering and monitoring are considered to be the major elements used in a smart grid. Devices are deployed, e.g., at the residential area for monitoring of the data, e.g., electricity, gas, water consumption, monitoring and controlling utility consumption, tracing, tracking and protecting people, and assets or animals. Following are the major use cases of smart metering and monitoring [67]:

Prepaid delivery of utilities Companies mainly deploy metering devices which are capable to send information to application servers either automatically or on request. For instance, M2M devices can be used to automatically bill the metering resource. Furthermore, another function of these devices is to improve energy efficiency and performance through accurately measuring readings of the resource consumption, e.g., water, gas, light, etc. Moreover, these applications aim to deliver data of the actual consumption without human intervention. Consequently, it provides and increases a positive environmental impact between service providers and their customers via autonomous communications. In addition, M2M modules can be placed in harsh environments. The limited space available for deploying M2M devices additionally demands small-sized modules. Moreover, the data should be secured and devices should be protected.

Pre-payment is one of the metering applications of gas, electricity, water, etc. With pre-payment, a person/family can buy a particular volume of gas, water, petrol, etc. This information is securely sent and stored within an M2M module. Furthermore, the actual and accurate amount of energy consumed during some particular time is sent to an M2M device. In addition, whenever the household consumes the acquired volume, the actual readings of the used energy are transmitted using an M2M module. When the limit of the purchased volume is reached, warnings in terms of messages or indicators can be sent and the supply can also be stopped.

Person, animal and object protection In these applications, the position as well as status of the person, animal and/or object are monitored. Portable M2M devices with communication modules and an optional GPS functionality are positioned on the body of a person, animal, and/or object. The collected data is sent to the monitoring/caring organizations either autonomously or on demand. The primary objective of these services is to improve the security of people, animals and objects through regular monitoring. Regarding the implementation part, these services are implemented through various applications running within the M2M device. Moreover, portable devices might undergo severe vibrations and shocks as these devices can be deployed in harsh and congested environments. Thus, the device size should be kept small enough to be placed in such an environment easily. In addition, protection of devices must be ensured to avoid cases of theft and misuse, as these devices store sensitive information.

1.6.4 POS terminals

Most of the existing POS terminals are connected through wired networks, and therefore fixed at some certain positions, e.g., in shopping malls, restaurants, super markets, railway stations, and airports. If a person is interested to use a POS terminal, then he has to visit the POS terminal. This might cause inconveniences for the users [67]. On the other hand, remotely located POS terminals (e.g., ticketing machines, and parking meters) require a wired connection which is not cost efficient and also difficult to be installed. In addition, this may also be exposed to damages. The other option is to take advantage of mobile networks to connect POS terminals. However, this approach poses some security limitations. Introducing M2M devices enable additional possibilities for applications. For instance, online transaction with credit/debit cards and secure communication can be offered with the help of M2M devices which can be deployed in, e.g., parking, stations, airports, and e-ticketing terminals. Typically, these devices have to fulfill several security requirements, e.g., for financial transactions.

1.6.5 M2M in control systems

M2M services provide improved intelligence in industrial plants, robots, furnaces, etc. Sensors and actuators are deployed which can collect and send information to the control room. For instance, if an emergency signal is detected by the sensors, it can be sent immediately to the relevant server. As a result, immediate remedial actions can be taken. Some of the major M2M applications in control systems are discussed below:

Controlling vending machines Due to the increasing product cost, diverse customer choice, increased demand for stock availability, and advanced regulations make a strong competition between operators of vending machines [132]. Vending machines are generally placed inside train stations, offices, public buildings, tourist spots, etc. Nowadays, people are recruited and dedicated to the maintenance of these vending machines [67]. They regularly visit the vending machines to perform several tasks such as checking the fill levels, re-fill if required, maintenance, as well as repairing damages. With the help of M2M technology, additional optimization possibilities are feasible, as a result vending machines can be managed and monitored in an efficient manner. By using devices connected with a mobile network, the current status such as details of the current stock, frequency of customers, damages, and malfunctions can be sent to monitoring centers, e.g., a quality control and maintenance team. In addition, devices can also send updates, e.g., about price variations and thus perform remote maintenance.

Controlling production machines Such machines are usually located inside the production sites which can be a harsh environment for production machines [67]. Dedicated people are hired to perform various tasks related to production machines, e.g., repair and maintenance. They visit production machines on a regular basis and perform several assignments such as repairing damages. The improved performance and maintenance of production machines can be achieved with the use of M2M communications. Thus, machines would be able to communicate with each other, and subsequently send updates to the respective control room. This can significantly increase visibility of the performance of production machines. Besides, in the case of damages or malfunctions, quick identification, and repairing tasks can be performed.

1.6.6 M2M compensation applications

M2M services in compensation applications include the following:

1.6.6.1 Utility account management

A customer will be able to receive an alert about his prepaid account in the case, e.g., when his prepaid account is expected to be finished soon [133]. Thus, he is expected to recharge his prepaid account, e.g., using online banking, and a mobile phone. Furthermore, a trusted third party such as a bank or a credit card company is usually involved in order to perform the verification as well as account recharging tasks. Furthermore, in the case of transportation and logistics, a driver can list, schedule, and deliver goods. Additionally, payments of the delivered goods can also be received electronically. Besides, enterprise POS (EPOS) applications and near-field communication (NFC) payments can be made by benefiting from the compensation capabilities.

1.6.6.2 Service primitives

In management, agreements and remunerations are the basic elements which each company has to deal with. ETSI in its Technical Specification TS 102 689 [133] specifies certain high level functions such as *CommitValue* and *Settle* which can be used to make agreements between two or more objects/entities. In addition, it is a prerequisite for the agreement that the object must have a legal account as well as relations with the trusted third party such as a bank or a credit card company.

1.6.7 M2M in home automation

M2M communications play a vital role in home automation to improve energy consumption, security, etc [134, 135]. One of the major focuses of home automation with M2M devices is to enhance energy consumption through regular metering and monitoring. For example, M2M devices can be deployed in a room to know about the presence of an individual. In the case, if there is no person present, the lights can be automatically switched off to save energy and ultimately to reduce costs. In addition, water and gas meters can be monitored and regulated using M2M devices. This will significantly improve energy consumption by increasing or decreasing the level of heat or water, e.g., in a room, office, stations, and shopping malls. M2M devices can be connected wirelessly to a gateway or directly to an eNB. Furthermore, the notion of heterogeneous technologies can also be used in such applications. It can receive data, merge with the other context information from other devices and send instructions to corresponding actuators.

Mobile M2M devices must use low-power consuming technologies in order to increase their lifetime. Consequently, the customer does not need to change batteries repeatedly. This is also needed for practical reasons, e.g., when M2M devices are installed in some places which are very difficult to approach. These applications can additionally help to inform customers about their energy consumption on a regular basis. This can be achieved with remote monitoring services, i.e., a customer/user can know about the amount of energy consumption even when the user is away from his home. This can also help to notify the user of any detected peculiarity (e.g., leakage of a gas). Consequently, immediate safety measures can be adopted.

1.7 What is coming next: MTC in 5G

The race towards developing next generation 5G networks has started. As a result an increasing interest is shown by several operators, automotive industry, as well as academia to deliver a general framework for 5G. One of the motivations behind the increasing interest in 5G is the expected thousand-fold rise in mobile traffic in 2020 compared to 2010 [136]. METIS is one of the first efforts worldwide to address the requirements for and expectations from this new generation 5G networks [82]. Furthermore, METIS aims to address upcoming challenges due to massive mobile traffic expected in 2020. The progress is achieved in several levels. Firstly, METIS particularly discusses five major 5G scenarios which include "Amazingly fast," "Great service in a crowd," "Ubiquitous things communicating," "Best experience follows you," and "Super real-time and reliable connections" [86, 137]. Furthermore, the major 5G objectives which are highlighted in the METIS project are given below [86]:

- Per area, the data volume is expected to increase by 1000 times.

- The number of supported devices is believed to be increased by 10–100 times.
- User data rates are anticipated to be increased by 10–100 times.
- In MTC devices with reduced power consumption, the battery life is expected to be increased by ten times.
- E2E delay is envisaged to be reduced by five times.

Supporting future MTC traffic is one of the major focuses of upcoming 5G networks [86]. In addition, 5G networks are expected to offer improved services in terms of data rates, latency, coverage, users/devices per cell, as well as power consumption for regular UEs and MTC devices. For these reasons, 5G networks are considered to be highly flexible and adaptive to cater the upcoming immense changes in the future. In addition to the aforementioned features, 5G networks are expected to provide high reliability and enhanced the quality of user experience (QoE) for both MTC and non-MTC traffic, therefore METIS has envisaged a user centric 5G network with multi-RAT[5] capabilities. In summary, the following services are expected for MTC in 5G networks and are depicted in Fig. 17 [137]:

1 xMBB (extreme mobile broadband): *xMBB* is capable to provide high data rate and at the same time improved QoE through moderate data rate provisioning. It can be seen from Fig. 17 that xMBB is stretched from gigabits per second to megabits per second with increasing reliability. This means that the xMBB provides high reliability for moderate data rates.

2 mMTC (massive MTC): The focus of 5G networks is to support low-cost and low-powered MTC devices used in several applications as well as co-exist with conventional non-MTC traffic. mMTC service is anticipated to support massive MTC devices with the reduced date rates when the number of connected sensors increases, as depicted in Fig. 17. Therefore, mMTC emphasizes to support a large number of devices.

3 uMTC (ultra-reliable MTC): 5G networks particularly focus MTC applications such as V2V, V2I, as well as industrial applications. Furthermore, these applications are expected to demand ultra-reliable and time critical services. Therefore, such applications demand high reliability. More importantly, services related to V2V/V2I communications demand additional fast communications in order to sense, deliver, receive, and react immediately. Therefore, uMTC emphasizes to support communication with increased reliability, as shown in Fig. 17.

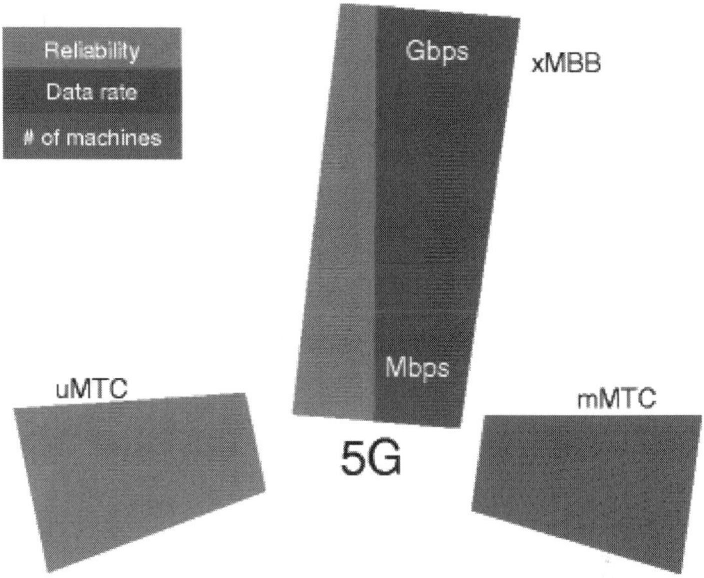

Figure 17. An overview of 5G service requirements for MTC communication (re-drawn from [137])

1.8 Open research areas

Most of the present M2M-related issues demand either modifications of the contemporary network architecture or standardization of a new network architecture especially for the RAN, to support future MTC in 3GPP systems [17]. Several operators as well as vendors are submitting proposals to 3GPP for the upcoming Release 13 and 14. Huawei along with Qualcomm and other major operators propose for a new network architecture to support mobile MTC. However, others such as Ericsson and Nokia are focusing to modify existing LTE network to support future MTC. Therefore, the system designers have to make sure whether an evolutionary or a revolutionary approach would be more effective for integrating future M2M traffic in mobile networks. In addition, from the perspective of standardization bodies, MTC enhancement tasks are ongoing in 3GPP Release 12 and upcoming Release 13 for optimizing mobile networks to serve MTC traffic. In addition, ETSI, IEEE, and other standardization organizations together with researchers from academia are putting efforts into developing and proposing standards for ensuring MTC communications efficiently. In this section, some of the following principal research issues which demand serious attention are discussed:

1.8.1 M2M traffic modeling

Due to the distinctive characteristics of M2M traffic, modeling the behavior of different kinds of M2M applications is one of the major requirements of M2M communications in the future. One of the possibilities is modeling each M2M device on its own. However, there are several challenges that need to be overcome. For instance, it is very difficult to model the behavior of traffic being generated from a massive amount of devices in parallel, and also it becomes more difficult in the presence of strong spatial and temporal correlation between devices [25]. On the other hand, aggregated traffic modeling defines M2M traffic as one stream from multiple devices. Nevertheless, the aggregated traffic models demand additional intermediate gateways between devices and eNBs. In [25], a coupled Markov modulated Poisson process (CMPP) framework was introduced, mainly targeting the source traffic modeling in M2M networks. Furthermore, the authors of [138] introduced the Beta/M/1 model for M2M communication considering beta-distributed inter-arrival time for the massive M2M traffic. However, at the same time, the authors stated that a G/M/1 queuing system may be more appropriate to model the behavior of different kinds of bursty applications in M2M communication. In addition, it was shown in [139], that Markov modulated Poisson process (MMPP) models are difficult to fit because they have a large number of parameters. Therefore, it is very difficult to solve them analytically. Since M2M traffic mainly exhibits periodic characteristics, stochastic models, e.g., the one given in [140] for modeling periodic traffic in an IP based system, can be developed to analyze the behavior of M2M traffic. Nevertheless, until now, there has been very little effort towards selecting the most suitable queueing models that can be used to represent the behavior of M2M traffic. In the future, there is a strong need for developing appropriate traffic models for the better understanding of M2M traffic characteristics.

1.8.2 Random access mechanism

As discussed in Subsection 1.5.3, the density of M2M devices per cell is expected to be enormous. Therefore, when these devices will try to access the network concurrently, this will lead to a low RA success rate and high network congestion in the PRACH. Consequently, this may cause unexpected delays, increasing packet loss, inefficient radio resource utilization, extra energy consumption as well as service intervention. Moreover, the random access channel can be further overloaded when these devices repeat their attempts to access the network after collisions. Thus, efficient access methods and overload control mechanisms are required to perform efficient RA procedures. Additionally, the initial access procedure in LTE-A networks consists of numerous steps which cause a significant amount of overhead and latency. Authors in [17] further stated that the state-of-the-art access procedures are not appropriate for M2M traffic. Either evolutionary (simpler) access schemes should be standardized for M2M communications or a revolutionary approach, i.e., fully separated standard LTE- and M2M-based data and control channels are required.

1.8.3 M2M battery life modeling

Power consumption of an M2M device is of significant importance to compare cellular- and non-cellular-assisted M2M services. Particularly, battery driven M2M devices demand low power consumption to increase their operating (life) time (as discussed in Subsection 1.5.6). Since M2M services are expected to be one of the potential sources to increase revenue of operators, it is important for module manufactures and service providers to design and ensure efficient mechanisms to reduce power consumption for M2M devices. One of the possible approaches can be that of investigating existing battery life models for UEs as well as M2M devices and designing an optimized battery life model based on the existing models and by considering more realistic M2M traffic characteristics.

1.8.4 M2M battery life optimization

Besides battery life modeling, optimized power consumption mechanisms can significantly increase operating time of an M2M device. For instance, extending sleep mode as well as longer TAUs can be effective to ensure longer battery lifetime. Moreover, optimized RRC, i.e., by optimizing signaling overhead can also be an efficient way to improve power consumption. In general, reducing idle mode activity can also be significantly helpful to improve battery lifetime.

1.8.5 Efficient PRB utilization

Existing mobile standards are particularly designed to support broadband traffic. Therefore, LTE TB sizes are only optimized for typical VoIP traffic. These mobile standards must be revised in order to support the enormous M2M traffic in the future. As mobile radio resources are a valuable asset, these limited radio resources must be efficiently utilized. Since the capacity of a PRB can be significantly larger than the need of typical M2M application (i.e., under favorable channel conditions), allocating 1 PRB to a single M2M device can notably degrade radio spectrum utilization. Hence, alternative schemes such as data aggregation and multiplexing are required for M2M communication. Aggregation and multiplexing of small M2M packets can be achieved by using an intermediate node between devices and eNB. Since relay nodes are one of the latest features of LTE-A networks, they can be used to aggregate traffic from multiple devices. In this way, small packets from different M2M applications can be added to make a comparatively larger packet to maximize radio resource utilization. Besides, it can significantly reduce the risk of network congestion due to the fact that not all devices in the vicinity of the relay node but only a relay node will communicate with the eNB.

2 CONCLUSIONS

LTE-A and upcoming 5G networks have particularly focused to dominate non-cellular technologies to support future M2M. Therefore, LTE-A is regarded as the ready-to-use technology for mobile M2M communication. One of the major challenges that mobile M2M poses is to support the myriad of devices sending small payloads, thus inefficiently utilizing PRBs, causing excessive signaling overhead, and increasing the risk of network congestion. Furthermore, provisioning of low-cost modules (up to US$10) and low-power devices is another challenge for manufacturers and service providers, respectively. To cater for these issues, schemes such as relay-based data aggregation can be applied to accommodate approximately 10 to 20 times more packets per PRB. Moreover, simpler and opportunistic RA procedures as well as cognitive M2M are candidates to be investigated with respect to reducing signaling overhead and to avoid network congestion. Besides, appropriate traffic models and battery life models need to be developed for future M2M applications.

Techniques for the performance evaluation of the preceding proposals should include all possibilities that are available, e.g., system level simulations, analytical modeling, and hybrid techniques, respectively. The scenarios to be investigated include mainly narrowband applications with small-sized packets, e.g., to evaluate the performance of data aggregation. However, broadband applications need to be part of the traffic mix and can also benefit from some of the changes, e.g., larger TBS. Similarly, the RA procedures are usually complex to model in simulators, thus are candidates to be modeled analytically. Moreover, the performance of simpler and opportunistic RA procedures can be determined and compared with the state-of-the-art RA procedures. Likewise, the performance of optimized power consumption schemes can be assessed through simulations as well as analytically. In addition, the performance of low-power M2M devices in terms of operating time can be determined analytically. Results can also be validated through test bed measurements. More importantly, the major application scenarios to be investigated for low power consumption include deployment of battery-driven devices in locations where there is no or limited direct power source.

Spectrum for mobile networks will remain a scarce resource also in the future and thus new concepts for new traffic types need to be carefully planned and evaluated. Additionally, the IoT will drive new developments including new devices and applications. The future thus poses challenges for all parties involved starting from standardization bodies, manufacturers of network equipment and user devices, network operators, and application providers. In the optimal case, research institutions support these developments by evaluating the new concepts in close cooperation.

3 ENDNOTES

[1] Machine-type communications (MTC) is identical to M2M communication in 3GPP documentations. Since this survey focuses on both ETSI and 3GPP standardization efforts, M2M and MTC terms are used interchangeably in the rest of the paper. Additionally, the terms M2M and MTC are used in the text on the basis of their conventional use by the respective standardization bodies. Moreover, the term non-M2M also called non-MTC represents traditional mobile H2H and H2M traffic.

[2] A bearer is defined as the flow of IP packets between gateways and UEs with a specific QoS.

[3] The SCS represents an entity which is capable to provide value-added services and performs user and control plane communication tasks for MTC [51].

[4] European Union's Seventh Framework Programme.

4 Multi random access technologies.

5

ACKNOWLEDGMENTS

We thank the International Graduate School for Dynamics in Logistics (IGS), doctoral training group of LogDynamics, University of Bremen, Germany, and EU programme Erasmus Mundus FUSION (Featured eUrope and South asIa mObility Network) Strand 1, Lot 11, Grant Agreement Reference number 2013-2541/001-011 EM Action 2 partnerships, for the financial support of this work.

COMPETING INTERESTS

The authors declare that they have no competing interests.

REFERENCES

1. ETSI, Machine-to-machine communications (M2M); Definitions (ETSI TR 102 725 V1.1.1, June, 2013).

2. T Taleb, A Kunz, Machine type communications in 3GPP networks: potential, challenges, and solutions. IEEE Commun. Mag. 50(3), 178–184 (2012).

3. Exalted: Expanding LTE for devices. Available from: http://www.ict-exalted.eu/fileadmin/documents/EXALTED_WP2_D2.1.pdf. Accessed 30 Sept 2013.

4. Ericsson, Towards 50 billion connected (2010). http://www.ericsson.com/au/res/region_RASO/docs/2010/ericsson_50_billion_paper.pdf. Accessed 10 January 2014.

5. ETSI, Machine-to-machine communications (M2M); use cases of Automotive applications in M2M capable networks (ETSI TR 102 898 V1.1.1, April, 2013).

6. NOKIA, LTE M2M: Optimizing LTE for the Internet of Things (Nokia white paper, 2014). Available from: file:///C:/Users/ yasir/ Downloads/nokia_lte-m_-optimizing_lte for_the_internet_of_things_white_paper.pdf. Accessed 5 August 2015.

7. COMARCH, Slicing up the M2M revenue pie: how to get your share and boost your business, (2010).

8. Y Chen, W Wang, in IEEE 72nd Vehicular Technology Conference Fall (VTC 2010-Fall). Machine-to-machine communication in LTE-A (IEEEOttawa, Canada, 2010), pp. 1–4.

9. ERICSSON, ERICSSON Mobility report on the pulse of the networked society, (2014). Available from: http://www.ericsson.com/res/docs/2015/ericsson-mobility-report-june-2015.pdf. Accessed 10 July 2015.

10. 3GPP LTE-Advanced. Available from: http://www.3gpp.org/technologies/keywords-acronyms/97-lte-advanced_links. html. Accessed 28 May 2015.

11. V Chandrasekhar, JG Andrews, A Gatherer, Femtocell networks: a survey. IEEE Commun. Mag. 46(9), 59–67 (2008).

12. 3GPP, Technical specification group radio access network; evolved universal terrestrial radio access (E-UTRA); relay architectures for E-UTRA (LTE-Advanced) (Release 9). 3GPP TR 36.806 V9.0.0, (2010).

13. C Cox, An introduction to LTE: LTE, LTE-Advanced, SAE and 4G mobile communications (Wiley, New York, 2012).

14. E Kartsakli, AS Lalos, A Antonopoulos, S Tennina, MD Renzo, L Alonso, C Verikoukis, A survey on M2M systems for mHealth: a wireless communications perspective. Sensors. 14(10), 18009–18052 (2014).

15. M Chen, J Wan, F Li, Machine-to-machine communications: architectures, standards and applications. KSII Trans. Internet Inf. Syst. 6(2), 480–497 (2012).

16. K Chen, S Lien, Machine-to-machine communications: technologies and challenges. Ad Hoc Networks (2013). In-Press.

17. AG Gotsis, AS Lioumpas, A Alexiou, M2M scheduling over LTE: challenges and new perspectives. IEEE Veh. Technol. Mag. 7(3), 34–39 (2012).

18. K Zheng, F Hu, W Wang, W Xiang, M Dohler, Radio resource allocation in LTE-advanced cellular networks with M2M communications. IEEE Commun. Mag. 50(7), 184–192 (2012).

19. M Hasan, E Hossain, Random access for machine-to-Machine communication in LTE-Advanced networks: issues and approaches. IEEE Commun. Mag. 51(6), 86–93 (2013).

20. ZM Fadlullah, MM Fouda, N Kato, A Takeuchi, N Iwasaki, Y Nozaki, Toward intelligent machine-to-machine communications in smart grid. IEEE Commun. Mag. 49(4), 60–65 (2011).

21. S Lien, K Chen, Y Lin, Toward ubiquitous massive accesses in 3GPP machine-to-machine communications. IEEE Commun. Mag. 49(4), 66–74 (2011).

22. Z Sheng, S Yang, Y Yu, AV Vasilakos, J Mccann, K Leung, A survey on the IETF protocol suite for the internet of things: standards, challenges, and opportunities. IEEE Wirel. Commun. 20(6), 91–98 (2013).

23. M Yang, Y Li, D Jin, L Zeng, X Wu, AV Vasilakos, Software-defined and virtualized future mobile and wireless networks: a survey. Mob. Networks and Appl. 20(1), 4–18 (2014).

24. J Wan, M Chen, F Xia, L Di, K Zhou, From machine-to-machine communications towards cyber-physical systems. Comput. Sci. Inf. Syst. 10(3), 1105–1128 (2013).

25. M Laner, P Svoboda, N Nikaein, M Rupp, in Proceedings of the Tenth International Symposium on Wireless Communication Systems (ISWCS). Traffic Models for Machine Type Communications (VDEIlmenau, Germany, 2013), pp. 1–5.

26. MZ Shafiq, L Ji, AX Liu, J Pang, J Wang, in ACM SIGMETRICS Performance Evaluation Review, 40. A first look at cellular machine-to-

machine traffic: large scale measurement and characterization, (2012), pp. 65–76.

27. 3GPP, Technical Specification Group Services and System Aspects; Service accessibility (Release 12) (3GPP TS 22.011 V13.1.0, 2014).

28. ETSI, Machine-to-Machine communications (M2M); Functional architecture (ETSI TS 102 690 V2.1.1, 2010). Available from: http://www.etsi.org/deliver/etsi_ts/102600_102699/102690/02.01.01_60/ts_102690v020101p.pdf.

29. A Kumar, J Sengupta, Y Liu, 3GPP LTE: the future of mobile broadband. Wirel. Pers. Commun.62(3), 671–686 (2012).

30. 3GPP, Technical Specification Group Radio Access Network; requirements for evolved UTRA (E-UTRA) and evolved UTRAN (E-UTRAN) (Release 9) (3GPP TR 25.913 V9.0.0, 2009).

31. 3GPP, Technical Specification Group Radio Access Network; evolved universal terrestrial radio access network (E-UTRAN); architecture description (Release 9) (3GPP TS 36.401 V10.0.0, 2010).

32. Innovations T, LTE in a nutshell: protocol architecture (2010). Available from: http://www.tsiwireless.com/docs/whitepapers/LTE. Accessed 2 August 2015.

33. A Meylan, 3GPP LTE Radio layer 2, RRC and Radio Access Network Architecture (3GPP TSG-RAN WG2, 2010). Available from: ftp://www.3gpp.org/Information/presentations/presentations_2010/2010_06_India/3GPP%20LTE%20Radio%20layer%202.pdf. Accessed 13 January 2015.

34. LTE E-UTRAN and its access side protocols. White Paper (2010). Available from: http://go.radisys.com/rs/radisys/images/paper-lte-eutran.pdf. Accessed 7 September 2015.

35. 3GPP, Technical Specification Group Radio Access Network; evolved universal terrestrial radio access (E-UTRA); radio resource control (RRC); protocol specification (Release 12) (3GPP TS 25.331 V12.6.0, 2015).

36. 3GPP, Technical Specification Group Radio Access Network; evolved universal terrestrial radio access (E-UTRA); packet data convergence protocol (PDCP) specification (Release 12) (3GPP TS 36.323 V12.2.0, 2014).

37. 3GPP, Technical Specification Group Radio Access Network; evolved universal terrestrial radio access (E-UTRA); radio link control (RLC) protocol specification (Release 12) (3GPP TS 36.322 V12.1.1, 2014).

38. 3GPP, Technical Specification Group Radio Access Network; evolved universal terrestrial radio access (E-UTRA); medium access control (MAC) protocol specification (Release 12) (3GPP TS 25.321 V12.2.0, 2014).

39. 3GPP, Technical Specification Group Radio Access Network; evolved universal terrestrial radio access (E-UTRA); LTE physical layer; general description (Release 12) (3GPP TS 36.201 V12.2.0, 2015).

40. 3GPP, Technical Specification Group Radio Access Network; evolved universal terrestrial radio access (E-UTRA); user equipment (UE) radio transmission and reception (Release 10) (3GPP TS 136.101 V10.3.0, 2011).

41. S Sesia, I Toufik, M Baker, LTE: the UMTS long term evolution (Wiley Online Library, New York, 2009).

42. 3GPP, Technical Specification Group Radio Access Network. Evolved universal terrestrial radio access (E-UTRA) and evolved universal terrestrial radio access network (E-UTRAN). Overall description. 3GPP TS 36.300 V11.2.0 (2012).

43. 3GPP, Technical Specification Group Radio Access Network. Evolved Universal Terrestrial Radio Access (E-UTRA) and Evolved Universal Terrestrial Radio Access Network (E-UTRAN). Overall description Stage 2 (Release 13). 3GPP TS 36.300 V13.0.0 (2015).

44. 3GPP, Technical Specification Group Radio Access Network; Evolved Universal Terrestrial Radio Access (E-UTRA); Medium Access Control (MAC) protocol specification (Release 12). 3GPP TS 36.321 V12.0.0 (2014).

45. 3GPP, Architecture enhancements to facilitate communications with packet data networks and applications (Release 12). 3GPP TS 23.682 V12.4.0 (2015).

46. 3GPP, Technical Specification Group Radio Access Network; Requirements for further advancements for Evolved Universal Terrestrial Radio Access (E-UTRA)(LTEAdvanced)(Release 12). 3GPP TR 36.913 V12.0.0 (2014).

47. P Mogensen, W Na, IZ Kovács, F Frederiksen, A Pokhariyal, KI Pedersen, K Kolding, K Hugl, M Kuusela, in IEEE 65th on Vehicular Technology Conference. LTE capacity compared to the shannon bound (IEEEDublin, Ireland, 2007), pp. 1234–1238.

48. U Barth, 3GPP Long-Term Evolution/system architecture evolution overview (Alcatel White Paper) (2006). Avilable from: http://www.ikr.uni-stuttgart.de/Content/itg/fg524/Meetings/2006-09-29-Ulm/01-3GPP_LTE-SAE_Overview_Sep06.pdf, accessed 19 February 2015.

49. ETSI, Machine-to-machine communications (M2M), Functional architecture (ETSI TS 102.690 V1.1.1, 2011). Avilable from: http://www.etsi.org/deliver/etsi_ts/102600_102699/102690/01.01.01_60/ts_102690v010101p.pdf, accessed 15 March 2015.

50. MJ Booysen, S Zeadally, GV Rooyen, Survey of media access control protocols for vehicular ad hoc networks. IET Commun. 5(11), 1619–1631 (2011).

51. 3GPP, Architecture enhancements to facilitate communications with packet data networks and applications (Release 13) (3GPP TS 23.682 V13.1.0, 2015).

52. 3GPP, Architecture enhancements to facilitate communications with packet data networks and applications (Release 11) (3GPP TS 23.682 V11.5.0, 2013).

53. 3GPP, Architecture enhancements to facilitate communications with packet data networks and applications (Release 13) (3GPP TS 23.682 V13.2.0, 2015).

54. 3GPP, Service requirements for machine-type communications (MTC) (Release 11) (3GPP TS 22.368 V11.5.0, 2012).

55. 3GPP, Technical Specification Group Services and System Aspects; service requirements for machine-type communications (MTC); Stage 1 (Release 13) (3GPP TS22.368 V13.1.0, 2014).

56. 3GPP, Technical Specification Group Services and System Aspects; Service requirements for machine-type communications (MTC); Stage 1 (Release 10) (3GPP TS 22.368 V10.5.0, 2011).

57. 3GPP, Technical Specification Group Services and System Aspects; Service requirements for machine-type communications (MTC); Stage 1 (Release 11) (3GPP TS 22.368 V11.6.0, 2011).

58. 3GPP, Technical Specification Group Services and System Aspects; Service requirements for machine-type communications (MTC); Stage 1 (Release 12) (3GPP TS 22.368 V12.4.0, 2014).

59. Z Fan, RJ Haines, P Kulkarni, M2M communications for E-health and smart grid: an industry and standard perspective. IEEE Wireless Communications. 21(1), 62–69 (2013).

60. Standards for M2M and the Internet of Things. Available from: http://www.onem2m.org/. Accessed 28 Aug 2015.

61. 3GPP, Study on enhancements for machine type communication (Release 12) (3GPP TR 22.888 V1.0.0, 2013).

62. 3GPP, Feasibility study on the security aspects of remote provisioning and change of subscription for machine to machine (M2M) equipment (TR 33.812 V9.1.0, 2010).

63. 3GPP, Security aspects of machine-type and other mobile data applications communications enhancements (Release 12) (3GPP TR 33.868 V0.10.0, 2012).

64. 3GPP, Standardization of machine-type communications, V0.2.4, (2014). Available from: http://wenku.baidu.com/view/163f99d5960590c69ec3767f. Accessed 01 July 2015.

65. ETSI, Machine-to-machine communications (M2M); Interworking between the M2M architecture and M2M area network technologies (ETSI TR 102 966 V1.1.1, 2014).

66. ETSI, Machine-to-machine communications (M2M); mIa, dIa and mId interfaces (ETSI TS 102 921 V1.1.1, 2012).

67. ETSI, Machine-to-machine communications (M2M); M2M service requirements (ETSI TS 102 689 V1.1.1, 2010).

68. ETSI, Machine-to-machine communications (M2M); Interoperability Test Specification for CoAP Binding of ETSI M2M Primitives (ETSI TS 103 104 V2.1.1, 2013).

69. ETSI, Machine-to-machine communications (M2M); use cases of M2M applications for e-health (ETSI TR 102 732 V1.1.1, 2013).

70. ETSI, Machine-to-machine communications (M2M); M2M service requirements (ETSI TR 102 691 V1.1.1, 2010).

71. ETSI, Machine-to-machine communications (M2M); use of automotive applications in M2M capable Networks (ETSI TR 102 898 V1.1.1, 2014).

72. ETSI, Machine-to-machine communications (M2M); applicability of M2M architecture in smart grid networks, impact of smart grid on M2M platform (ETSI TR 102 935 V2.1.1, 2012).

73. IEEE 802 16p-11/0033, Enhancements to support machine-to-machine applications for WirelessMAN-Advanced (IEEE 802.16's Machine-to-Machine (M2M) Task Group, 2010). Available from: http://ieee802.org/16/m2m/. Accessed 03 July 2015.

74. IEEE 802 16p-10/0004, Machine to machine (M2M) system requirements document (SRD) (IEEE 802.16Šs Machine-to-Machine (M2M) Task Group, 2010). Available from: http://ieee802.org/16/m2m/. Accessed 09 September 2014.

75. G Jones, OMA and Machine-to-Machine (M2M) communication (White Paper), (2011). Available from: http://openmobilealliance.org/static/oma-annual-reports/documents/oma%20collateral%20m2m%205-11.pdf. Accessed 17 August 2015.

76. TIA TR-50 framework intelligent device/smart devices. Available from: http:// ftp.tiaonline.org/TR-30/TR-30.2/Public/2009%20Meetings/ 2009-09%20Arlington/20909008b%20TIA%20Intelligent%20Device%20 Persentation.pdf. Accessed 28 Aug 2015.

77. Remote SIM Provisioning for machine to machine. Available from: http://www.gsma.com/connectedliving/embedded-sim/. Accessed: 12 June 2015.

78. AT, 4 wireless laboratories announces the first NFC Forum device certification. Available from: http://www.m2mnow.biz/2012/03/31/4833-at4-wireless-laboratories-announces-the-first-nfc-forum-device-certification/. Accessed 20 July 2015.

79. Discover Wi Fi. Available from: http://www.wi-fi.org/discover-wi-fi. Accessed 12 June 2015.

80. Open Mobile Alliance (OMA). Available from: http:// openmobilealliance.org/about-oma/work-program/m2m-enablers/. Accessed 01 Apr 2014.

81. China Communications Standards Association. Available from: http://www.ccsa.org.cn/english/. Accessed 20 July 2015.

82. Mobile and Wireless Communications Enablers for Twenty-Twenty (2020). Available from: https://www.metis2020.com/?doing_ wp_cron= 1440943491553276062011718750000_links.html. Accessed 28 Aug 2015.

83. Mobile and wireless communications Enablers for Twenty-twenty (2020) Information Society, Novel radio link concepts and state of the art analysis. Deliverable Number ICT-317669-METIS/D2.2, 2013. Available from: https://www.metis2020.com/wp-content/uploads/deliverables/METIS_D2.2_v1.pdf, accessed 12 October 2014.

84. Achieving low-latency in wireless communications (LOLA). Available from: http://www.ict-lola.eu/. Accessed 30 June 2015.

85. CISCO, Cisco Visual Networking Index: Global Mobile Data Traffic Forecast Update. 2014–2019 (white paper), http:// www.cisco.com/c/en/us/solutions/collateral/service-provider/ visual-networking-index-vni/white_paper_c11-520862.pdf Accessed 20 July 2015.

86. Mobile and wireless communications Enablers for Twenty-twenty (2020) Information Society, Scenarios, requirements and KPIs for 5G mobile and wireless system (Deliverable Number ICT-317669-METIS/D1.1, 2013). Available from: https://www.metis2020.com/wp-content/uploads/ deliverables/METIS_D1.1_v1.pdf, accessed 12 October 2014.

87. Cellular M2M forecasts: unlocking growth. Available from: https:// gsmaintelligence.com/research/2015/02/ cellular-m2m-forecasts-unlocking-

growth/457//%20Accessed:%206%20 Feb%202015/_links.html. Accessed 23 May 2015.

88. Report: M2M device connections forecast 2014-2024. Available from: http://www.machinetomachinemagazine.com/2014/08/20/report-m2m-device-connections-forecast-2014-2024/. Accessed 23 May 2015.

89. T Pötsch, SNK Marwat, Y Zaki, C Görg, in IEEE 6th Joint IFIP Wireless and Mobile Networking Conference (WMNC). Influence of Future M2M Communication on the LTE system (IEEE,Dubai, UAE, 2013), pp. 1–4.

90. Y Mehmood, SNK Marwat, T Pötsch, F Ahmad, C Görg, I Rashid, Impact of M2M traffic on LTE data traffic performance (Springer, Bremen, Germany, 2014).

91. P Kela, J Puttonen, N Kolehmainen, T Ristaniemi, T Henttonen, M Moisio, in 3rd International Symposium on Wireless Pervasive Computing. Dynamic packet scheduling performance in UTRA long term evolution downlink (IEEESantorini, Greece, 2008), pp. 308–313.

92. G Monghal, KI Pedersen, PE Mogensen, in IEEE Vehicular Technology Conference. QoS oriented time and frequency domain packet schedulers for the UTRAN long term evolution (IEEESingapore, 2008), pp. 2532–2536.

93. Y Niu, C Gao, Y Li, L Su, D Jin, A Vasilakos, Exploiting Device-to-Device Communications in Joint Scheduling of Access and Backhaul for mmWave Small Cells, (2015). accepted in IEEE Journal on Selected Areas in Communications http://arxiv.org/pdf/1503.02292v1.pdf.

94. M Ding, D López-Pérez, AV Vasilakos, W Chen, in IEEE International Conference on Communications Workshops (ICC). Dynamic TDD transmissions in homogeneous small cell networks (IEEE,Sydney, NSW, 2014), pp. 616–621.

95. M Ding, D Lopez-Perez, X Ruiqi, AV Vasilakos, W Chen, in IEEE International Conference on Communications (ICC). Small cell dynamic TDD transmissions in heterogeneous networks (IEEE,Sydney, NSW, 2014), pp. 4881–4887.

96. M Ding, D López-Pérez, AV Vasilakos, W Chen, in IEEE Global Communications Conference (GLOBECOM). Analysis on the SINR performance of dynamic TDD in homogeneous small cell networks (IEEE,Austin, Texas, 2014), pp. 1552–1558.

97. CY Wang, CH Ko, HY Wei, AV Vasilakos, A voting-based femtocell downlink cell-breathing control mechanism. IEEE/ACM Trans. Networking (TON), (2014).

98. MA Khan, H Tembine, AV Vasilakos, Game dynamics and cost of learning in heterogeneous 4G networks. IEEE J. Sel. Areas Commun. 30(1), 198–213 (2012).

99. D López-Pérez, X Chu, AV Vasilakos, H Claussen, On distributed and coordinated resource allocation for interference mitigation in self-organizing LTE networks. IEEE/ACM Transactions on Networking (TON). 21(4), 1145–1158 (2013).

100. D López-Pérez, X Chu, AV Vasilakos, H Claussen, Power minimization based resource allocation for interference mitigation in OFDMA femtocell networks. IEEE J. Sel. Areas Commun. 32(2), 333–344 (2014).

101. AS Lioumpas, A Alexiou, in IEEE GLOBECOM Workshops (GC workshops). Uplink scheduling for machine-to-machine communications in LTE-based cellular systems (IEEE,Houston, Texas, 2011), pp. 353–357.

102. S Lien, K Chen, Massive access management for QoS guarantees in 3GPP machine-to-machine communications. IEEE Commun. Lett. 15(3), 311–313 (2011).

103. AG Gotsis, AS Lioumpas, A Alexiou, Analytical modelling and performance evaluation of realistic time-controlled M2M scheduling over LTE cellular networks. Trans. Emerg. Telecommun. Technol. 24(4), 378–388 (2013).

CHAPTER 7

An Average Power Reduction Method for Web Applications on Wireless Terminals

Toshiya Koyasu, Hideki Shimada, Kenya Sato

Department of Information Systems Design, Doshisha University, Kyoto, Japan

ABSTRACT

Recently, there is a widespread use of Web browser-based Web applications such as e-mails and chats. However, frequent communication between mobile wireless terminals and HTTP servers give rise to problematic increases of average power consumption for the mobile devices. Current attempts to tackle this problem focus on reducing power consumption at the network and data link layers used by the devices. In this study, we propose a different solution where certain functionalities of the mobile application are delegated to other devices with abundant power resources (either from large capacity batteries or power outlets). This method off-loads parts of the communication process normally done at wireless terminals to amply powered machines. With messages pushed from the machines only when the HTTP server responds with an update, the wireless terminal needs to transmit data less frequently, thus cutting down on power consumption. Our prototype implementing the proposed method succeeded in reducing the rise in average power consumption.

KEYWORDS

Web Applications; Ajax; Wireless Communication

1. INTRODUCTION

Recent years have seen a growth of smart phones and other wireless devices with sophisticated Web browsers [1], making it possible and more practical to use Web pages and applications from wireless terminals. This has also made common the use of Ajax applications from wireless terminals. Ajax applications are notable for their high usability and their convenience of not having to install additional software. Some examples of these are text-based chats and email applications that let users communicate using the Web browser as the interface.

However, the use of Web applications with Ajax technology on a wireless terminal, ones that synchronize data updates on the server with Web browser displays, poses a problem: High power consumption. In this study, we propose a method of reducing power consumption at the wireless terminal while using Ajax [2] applications.

2. DEVICE POWER CONSUMPTION

2.1. Web Applications

Ajax is an approach to Web applications that attempts to improve on usability. Web application usability is tied to the way the browser and the HTTP server communicate with each other: a critical functionality for a Web application. Traditional Web applications without Ajax would reload the whole displayed page every time this needs to be done. This action causes unfavorable user experience because it temporarily removes UI components such as text boxes and buttons from the display, and could possibly interrupt the user's tasks. On the other hand, Ajax applications only update parts of the display, only when they need to be updated. UI components remain visible, which leads to higher usability.

Figure 1(a) shows an example of how an Ajax application works. In this application, the user types a product ID in the ID text box and presses the SEARCH button. A search is then conducted in the remote database for the corresponding product. The result is displayed in the NAME text box while other parts of the UI remain as they are. Figure 1(b) is the sequence of actions beginning with the ID input and ending with the product name display.

- Web browser: The front-end of the Web application.

- HTTP Server: A collection of Web, application and database servers that is the back-end of the Web application.
- XHTML File: The file that defines the visuals and functionalities of the Web application.
- Javascript: The scripting language used to embed extended functionalities on the Web page.
- Ajax engine: The Javascript program (a group of functions) that provides HTTP communication and UI updating.
- XML Http Request: An HTTP client library that can be used from a Javascript program.
- UI Tags: The XHTML tags that define the Web application's UI.

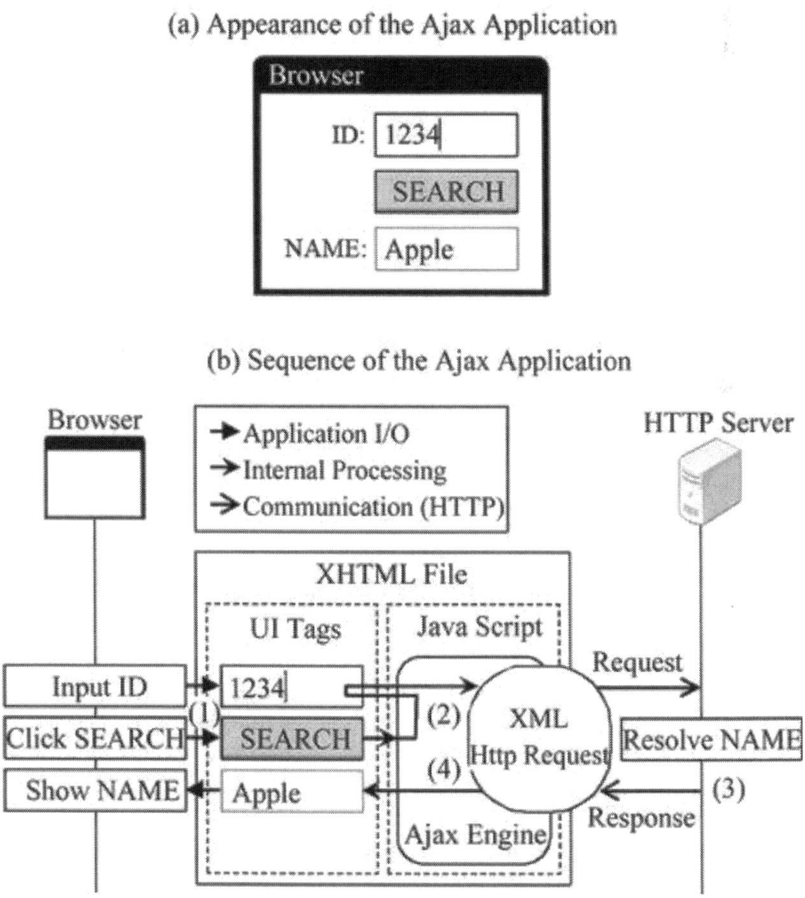

Figure 1. Behavior of the Ajax application.

The basic components and actions 1) through 4) in the figure are as follows:

1 The user types the product number in the ID text box and presses the SEARCH button. A function defined in the Ajax engine is called as an event handler for the button press event.

2 The function retrieves the ID text box value and sends it to the server using XML Http Request. At the same time, another Ajax engine function is registered as a handler for the event that occurs when a response is received.

3 The HTTP Server resolves the product name based on the value received, and sends back a response to XML Http Request.

4 XML Http Request calls the event handler registered in action 3). The handler inserts the product name into the NAME text box.

2.2. Synchronous Applications

An application where browser display updates are synchronized with server data updates, such as a browserbased chat application, would update the browser display to show a newly submitted message when it is posted to the HTTP server. We call this type of application a "synchronous application". The Web being a pull-based communication system, a synchronous application would need to somehow send request messages periodically from the browser to the HTTP server. We will take the chat application as an example to describe how these messages can be sent.

1) Non-Ajax method

B-Chat [3] achieves periodical requests with client pull, a feature in which XHTML files are re-retrieved and redisplayed periodically. B-Chat uses this feature to get the XHTML file with the latest messages every once in a while. With client pull, the whole browser display must be reloaded every time synchronization occurs, so the user wouldn't be able to operate on the page if the sync interval is too short. Therefore, applications that use client pull tend to set the sync interval to longer periods compared applications using Ajax. B-Chat uses a 30 second sync interval.

2) Ajax method

Ajax Chat [4] achieves periodical requests using Javascript functions in its Ajax engine. Applications using this method can set shorter sync intervals because only specific parts of the browser display needs to be updated. The user is not interrupted by synchronization. Rather, the shorter the sync interval, the smaller the latency of display update after data is updated on the server, so applications that use Ajax tend to set shorter sync intervals compared non-Ajax applications. Ajax Chat uses a 2 second sync interval.

Mobile wireless terminals consume more power when communicating via its wireless network interface. A study [5] done by Carroll et al. shows that a particular device's power consumption while communicating was 1016.4 mW, which was 6.31 times higher than while off communication (161.2 mW). We can generally say that using an application with a high rate of communication on a wireless terminal increases its energy consumption. A synchronous application is one of them. Ajax applications in particular communicate more frequently compared to others, so the problem is grave and needs to be addressed.

3. PROPOSED METHOD

Approach

To reduce average power consumption on a device using synchronous Ajax applications, we propose the two following approaches to off-load transactions and reduce communication frequency.

- Delegating communication to other machines:

We use other devices to carry out the actual communications for the wireless terminal while data updates are less frequent. We call this other device the "Surrogate Device". Devices fit for this role are ones with network connections that have large capacity batteries or enjoy access to steady power from outlets. Some examples would be desktop PCs with their screen savers running, idle servers, and networked home appliances [6] that are to be introduced in the near future.

- Using push-based communications to deliver updates:

We use push-based communications to send updated data to the wireless terminal after the Surrogate Device retrieves them from the HTTP server. In pushbased communications, the data is sent from the publisher to the recipient without the recipient requesting it. Thus the wireless terminal does not need to poll the Surrogate Device for updates.

Figure 2 shows the proposed communication model for this transaction off-loading. Descriptions of the communications 1) through 4) are as follows:

1 Direct communication: This phase is used to detect communication done by synchronous Ajax applications against the HTTP server. After a communication is detected, the system checks to see if there are updates at the HTTP server. If the data is updated frequently at the remote server, the wireless terminal continues to communicate directly with the server, as off-loading would only have trivial effects. This also avoids the overhead time produced by off-loading.The same is done for non-Ajax synchronous applications; they communicate directly with the HTTP server.

2 Surrogate Communication Request: If data updates at the HTTP server are infrequent, then the wireless terminal requests the Surrogate Device to carry out the actual communications. We call this request the "Surrogate Communication Request", and the message sent at this time the "Surrogate Communication Request Message". The wireless terminal suspends communications after sending out this request.

3 Surrogate communication: The Surrogate Device, upon receiving a Surrogate Communication Request, checks the HTTP server for data updates.

4 Surrogate Communication Response: If data is updated on the remote server, the Surrogate Device ends communications with the HTTP server and pushes the updated data back to the wireless terminal. We call this push the "Surrogate Communication Response" and the message sent at this time the "Surrogate Communication Response Message".

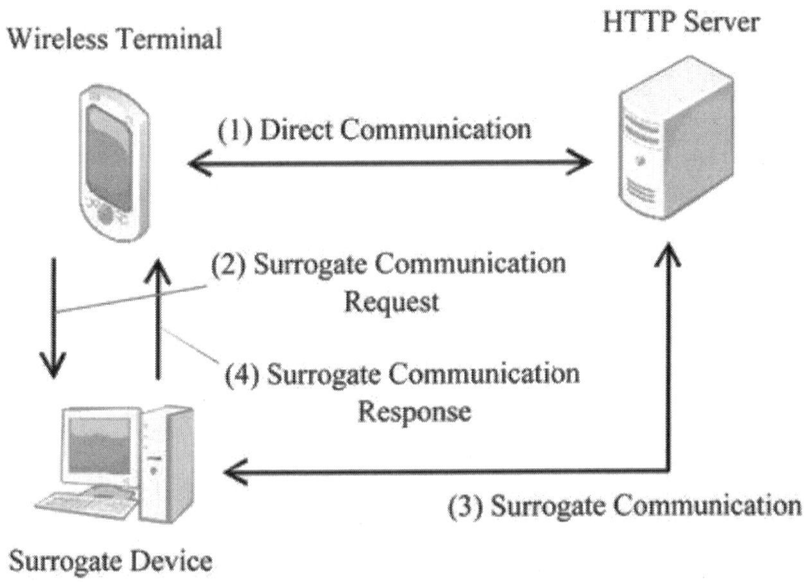

Figure 2. Communication model of the transaction offload.

4. PROTOTYPE IMPLEMENTATION

Figure 3 shows the setup of the prototype we implemented to evaluate the proposed method. Each of the components are shown corresponding to Figure 2. The prototype consists of the Master Gateway (MGW), residing on the wireless terminal, and the Slave Gateway (SGW), residing on the Surrogate Device.

4.1. Master Gateway

The MGW is the module to be used by the browser as a local proxy server. As with a generic local proxy server, the MGW relays communications between the browser and the HTTP server and caches the server responses. In addition to this basic functionality, it also distinguishes the transactions that are taking place. This is done based on the request URI after stripping it of the query parameters. Each of the transactions is assigned a defined state. The state assigned to a transaction is checked when the browser sends a request. Subsequent behavior of the MGW depends on the state of each of the transactions. The four states and their corresponding actions are as follows:

- Ajax Communication Detection State AJAX_DET:

In the initial state, AJAX_DET, the MGW scans the communications between the browser and the HTTP server and detects those that are done by Ajax applications. This is done by finding JSON objects (packages of data often used with Ajax applications) in the HTTP server's response messages. The MGW finds JSON objects by checking the Content-Type header in HTTP responses. Once the MGW detects an Ajax communication, the state transitions to UPSTOP_ DET.

- Update Stop Detection State (UPSTOP_DET):

In the UPSTOP_DET, the MGW attempts to detect whether updates at the HTTP server has ceased temporarily. The MGW decides that updates have ceased temporarily when the latest response from the HTTP server is identical to the most recently cached response. Once a temporary cease in communications has been detected, the MGW sends a request message to the SGW (the Surrogate Communication Request Message) that contains both the browser's request and the server's response, combined as a multi-part message body. This is the Surrogate Communication Request Message. The state now transitions to RES_WAIT.

- Response Waiting State (RES_WAIT):

In RES_WAIT, the MGW waits for a response from the SGW. Any Ajax requests sent from the browser are responded with cached contents. Completing all communications within the wireless terminal in this way has the effect of practically suspending its communications. Once the SGW responds with a Surrogate Communication Response Message, the cache is overwritten with the newly arrived data. The state now transitions to RESD.

- Surrogate Response Delivery State (RESD):

In RESD, the MGW delivers the Surrogate Communication Response Message to the browser. The message, which is the cache contents that have been overwritten, is delivered in response to requests from the browser. The state transitions back to UPSTOP_DET at this point.

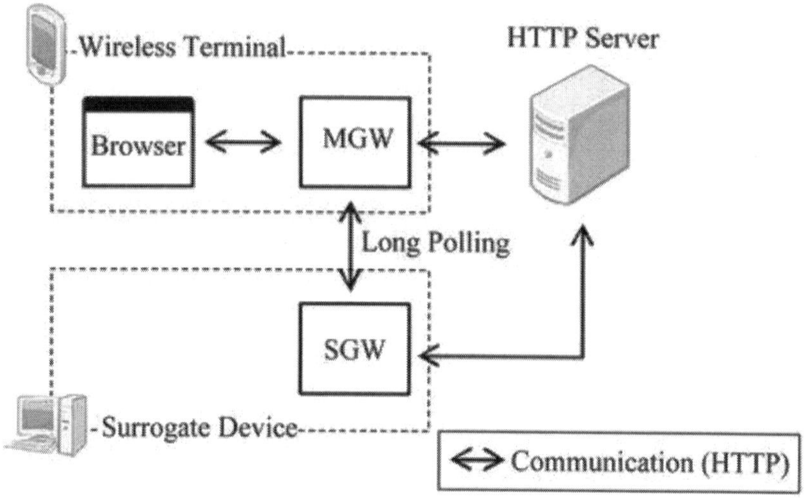

Figure 3. Components of the prototype.

4.2. Slave Gateway

The SGW is the module that receives Surrogate Communication Requests from the MGW and carries out the communications for the wireless terminal; it sends out the request message contained in the Surrogate Communication Request Message periodically to the HTTP server. Consecutive responses from the server are checked to see if they differ from the sample response message contained in the Surrogate Communication Request Message. If in fact the responses do differ, the SGW decides that the remote data has been updated, and pushes the new response (the Surrogate Communication Request Message) to the MGW and terminates communications with the server.

4.3. Method of Content Transmission

For content transmission from the SGW to the MGW, we adopt the long polling method, which is used to achieve pseudo-pushing in pull-based communication systems such as the Web. In long polling, the client initially sends a request to the server. The server, instead of responding instantly, holds the request until a certain event occurs (usually the arrival of newly available data). Once this event occurs, the server sends a response to the client.

In this prototype, the SGW holds the Surrogate Communication Request until it detects data updates at the server.

5. EVALUATIONS

5.1. Evaluation Items

To evaluate the proposed method, we implemented and ran a sample application (described later) in various situations for 10 minutes each and evaluated the following two points.

- The Number of transmissions by the wireless terminal: We compared the number of times the wireless interface was used to send and receive messages.
- Average latency: We compared the average latency of the browser display update since the data update on the server.

5.2. Evaluation Method

We used an application resembling the communication model of an Ajax synchronous application, the chat. This application has the following two front-ends.

1) Sender

The Sender front-end is the one that submits new chat messages. It submits messages at normally distributed random time intervals having an average of μ (msec.) and a standard deviation of 250 (msec.). The requests have two parts to its query parameter. One is a timestamp of the request (the number of milliseconds since January 1st, 1970 at 00:00) that represents the time of submission of the chat message. The other is a sequence of 140 randomly selected Japanese Hiragana characters, encoded in UTF-8, that represents the actual chat message.

2) Receiver

The Receiver front-end is the one that retrieves submitted chat messages. It sends out a request to the HTTP server once every 3 seconds. The request has the timestamp of the request as the query (in the same format as the Sender). The HTTP server responds with chat messages that were submitted later than this timestamp.

5.3. Method of Content Transmission

Table 1 shows the combinations of evaluation conditions, consisting of μ (the average time intervals between message submissions by the Sender) and T (the time intervals between surrogate communications by the surrogate communication device). T is shown as “n/a” where this method is unnecessary.

Table 1. Evaluation cases.

	case 1	case 2	case 3	case 4
μ (msec)	3000	3000	9000	9000
T (msec)	n/a	1500	n/a	1500
	case 5	case 6	case 7	case 8
μ (msec)	9000	27000	27000	27000
T (msec)	3000	n/a	1500	3000

5.4. Method of Content Transmission

Specifications of each device used in the evaluation are shown in Tables 2-4. The devices all reside in different networks, and the Sender and Receiver are used on the wireless terminal.

During evaluation, we turned off the wireless device’s HSDPA interface and used its IEEE 802.11 g interface, and limited its throughput to that of HSDPA, which we had measured in advance. This is in consideration of the possibility that the HSDPA interface’s unstable throughput may cause unpredictable effects on evaluation. We used Iperf 1.7.0 [7] with a window size of 32 KB to measure the HSDPA interface’s through put. The limit applied to the IEEE 802.11 g interface is the average of 20 transmissions between the wireless terminal and the HTTP server. The actual values are 80 Kbps upstream and 40 Kbps downstream. We used NEGiES 1.57 [8] to apply the throughput limits.

5.5. Evaluation Procedure

First, the Receiver starts retrieving chat messages. Once the first response from the HTTP server is retrieved, the Sender starts submitting chat messages. Both front-ends are stopped after 10 minutes (counting from the initial message retrieval by the Receiver). The number of transmissions by the wireless terminal and the average latency are then evaluated.

Table 2. Specification of the wireless terminal.

CPU	Atom Z520 1.33 GHz
Memory	1.00 GB
Network	HSDPA [9], IEEE 802.11 g
Web Browser	Firefox 3.6.13

Table 3. Specification of the surrogate device.

CPU	Core2 Duo U9400 1.40 GHz
Memory	3.00 GB
Network	100 BASE-TX

Table 4. Specification of the HTTP server.

CPU	Core 2 Duo E7500 2.93 GHz
Memory	2.00 GB
Network	100 BASE-TX

5.6. Evaluation Results

The resulting number of communications and the average latency are shown as follows: cases 1 and 2 in Figures 4 and 5, cases 3 through 5 in Figures 6 and 7, cases 6 through 8 in Figures 8 and 9.

6. DISCUSSION

Let us define the average increase in the wireless terminal's power consumption, compared to when the terminal's communications are halted, as

$$w_{avg} = \left(W_{com} - 1\right)\frac{t_{com}}{t} \qquad (1)$$

Here, W_{com} is the rate in which power consumption increases or decreases when communicating, compared to when it is not. We set this according to the study by Carroll et al. (cited in Chapter 2) to

$$W_{com} = 6.31 \tag{2}$$

it is the length of time the sample application ran, which is

$$t_{com} = 10 \times 60 = 600 \tag{3}$$

t_{com} is the total length of time (sec) the Receiver used the wireless terminal's interface to communicate.

Figure 10 shows the values of Equation (1) when applied to the evaluation results for cases 3 through 5. Figure 11 is for cases 6 through 8.

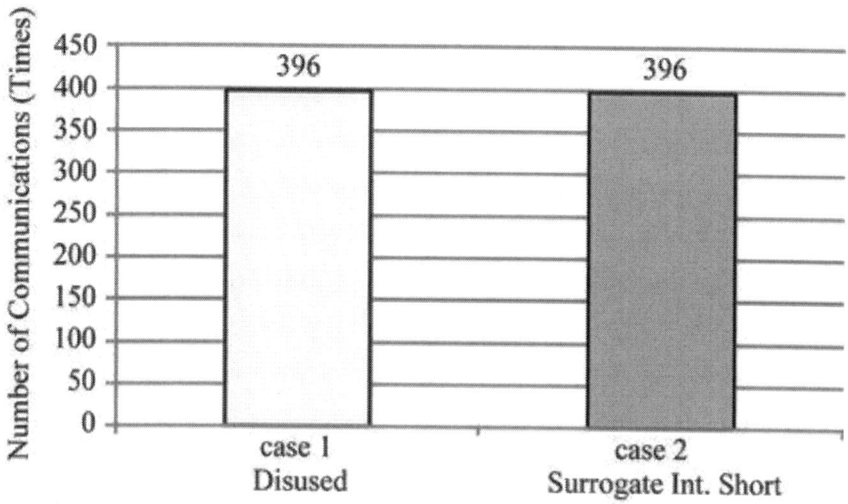

Figure 4. Number of communications on cases 1 and 2.

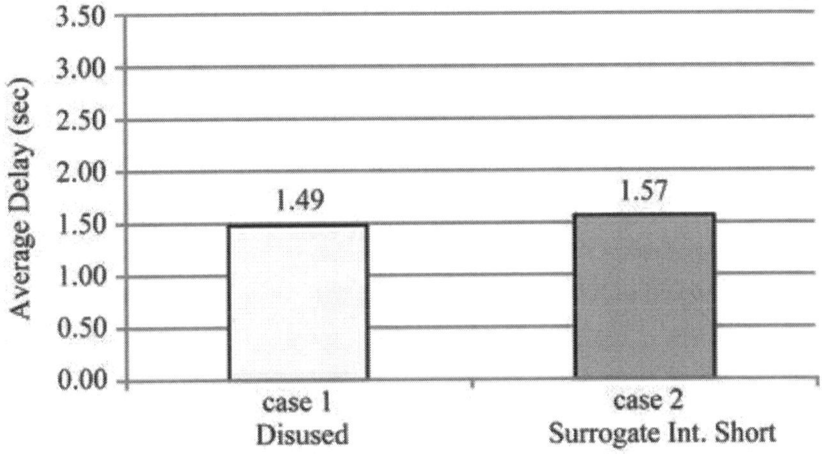

Figure 5. Average delay on cases 1 and 2.

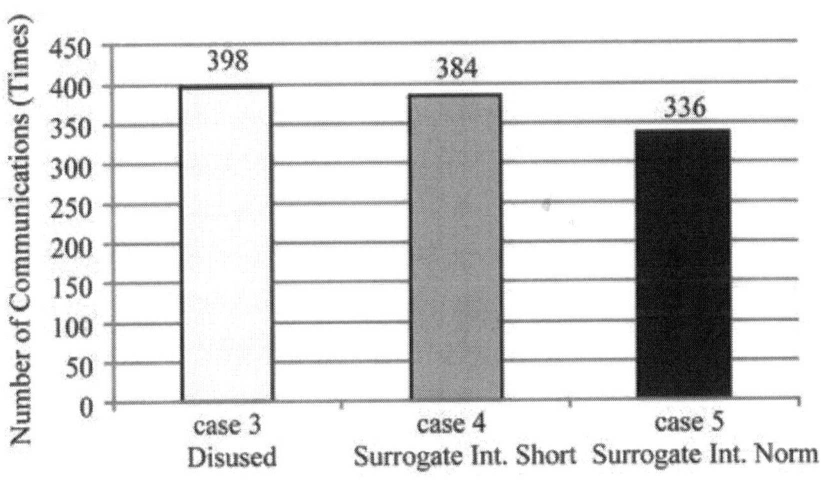

Figure 6. Number of communications on cases 3 to 5.

From Figure 11, we can see that the wireless terminal consumed 14% more power when using the sample application (case 6) and that our proposed method lowered it to 7% (cases 7 and 8). However, in Figure 10, the average power consumption when not using our method is 19% (case 3), which is lower than the 20% when it was used (case 4).

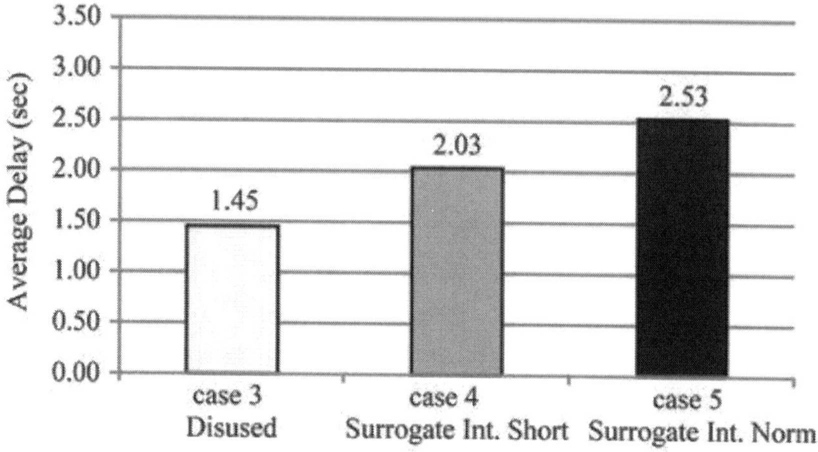

Figure 7. Average delay on case 3 to 5.

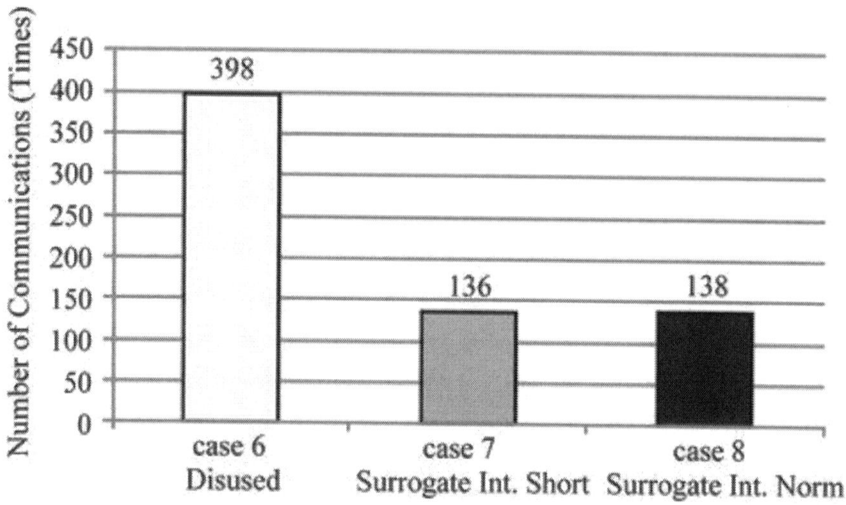

Figure 8. Number of communications on cases 6 to 8.

There are two possible reasons for this. One is that the HTTP server's data update rate (approximately once every 9 seconds) was relatively low for the sync rate (approximately once every 3 seconds). As such, the number of transmissions eliminated was a trivial 14 times. The other is that the size of the Surrogate Communication Request Message (963 bytes) was larger than that of the data update request message in long polling (434 bytes) and the wireless terminal had to communicate for a longer total time.

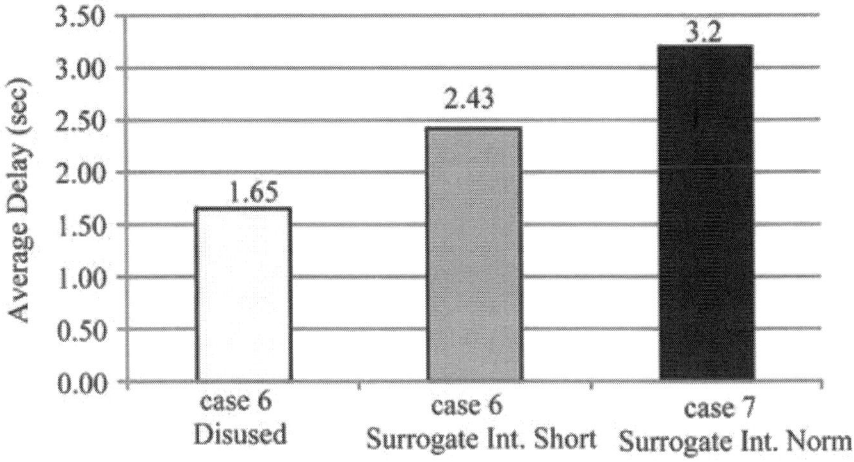

Figure 9. Average delay on cases 6 to 8.

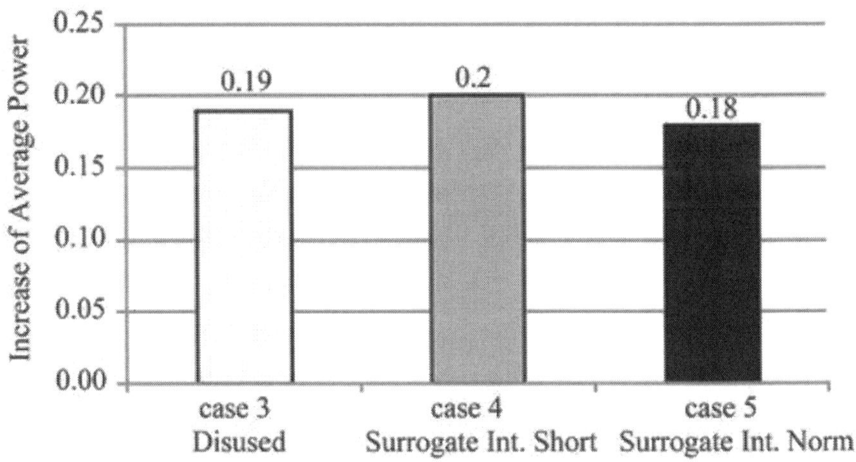

Figure 10. Increase of average power on cases 3 to 5.

In summary, the proposed method can reduce power consumption given that data update intervals are sufficiently long compared to the sync interval of the synchronous application. More specifically, we can conclude that our method is effective if the data update intervals are more than 9 times longer than the sync interval.

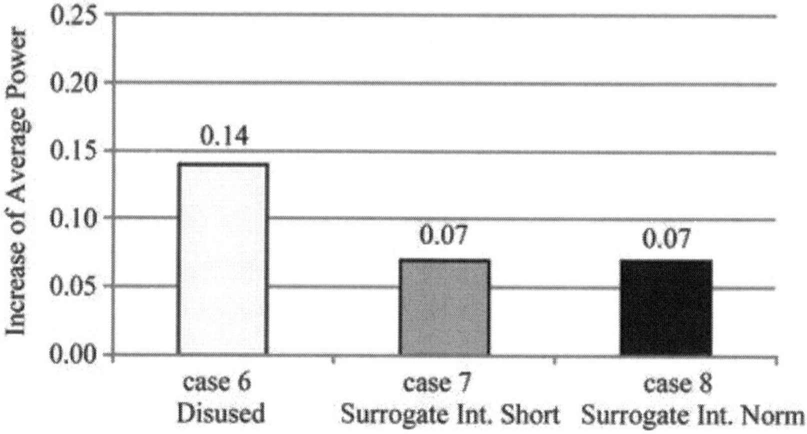

Figure 11. Increase of average power on cases 6 to 8.

6.1. Overhead Time

Figure 5 shows that when surrogate communications do not occur (case 2), our method has an increased average overhead time of under 100 ms compared to when our method is not used (case 1), which is acceptable. Figures 7 and 9 show that when surrogate communications do occur (cases 4, 5, 7 and 8), the increased average overhead time can be held down to under 1 second if the intervals of surrogate communications are approximately half of the sync interval.

With the above evaluation results, combined with the fact that synchronous applications do not impose realtime constraints, we conclude that the overhead time introduced by our proposed method does not impair practicality.

6.2. Comparisons with Related Methods

In the method proposed by Ishida et al. [10], the wireless interface on the terminal is set to sleep mode every time a transmission ends. A device called the Wakeup Module is then used in place of the wireless interface to wait for incoming transmission waves and wake up the wireless interface when needed.

In the method proposed by Russel [11], the front-end and the back-end of the Ajax application are modified to adapt long polling (described in Section 4.3) to reduce the communication rate: The HTTP server waits until there is new data available to respond to the browser.

Both of these methods require specific expertise in mobile devices or Web applications if they are to be applied to existing applications. As such, it is difficult if not impossible for the end user to use those methods. In contrast, our method has the advantage of being able to be used by the end user, since it does not require that modifications be made on existing applications.

7. CONCLUSIONS

In this study, we proposed a method of off-loading a wireless terminal's communications to another device in order to tackle the problem of increased power consumption that occurs while using synchronous Ajax applications (applications that synchronize the Web browser's display with data updates on the server). Our evaluation prototype kept the average increase in power consumption.

We used a networked device as the Surrogate Device in this particular study, but this method could potentially be adapted to wireless base stations and Web servers themselves.

Our method is a valuable contribution in the application software level to the wide social needs [12] of longer battery lives. Because this is a high-level method, we can expect it to be used in conjunction with other low-level technologies in hardware and communications.

ACKNOWLEDGEMENTS

This work was supported by JSPS Grant-in-Aid for Scientific Research KAKENHI (2100084).

REFERENCES

1. M. Stanley, "The Mobile Internet Report," 2009. http://www.morganstanley.com/institutional/techresearch/mobile_internet_report122009.html

2. J. J. Garrett, "Ajax: A New Approach to Web Applications," 2005. http://www.adaptivepath.com/ideas/ajax-new-approach-web-applications

3. http://www.adaptivepath.com/ideas/essays/archives/000385.php

4. http://wws.cside.com/cgi-plant/b_chat/blueimp.net, https://blueimp.net/ajax/

5. A. Carroll and G. Heiser, "An Analysis of Power Consumption in a Smart phone," Proceedings of the 2010 USENIX Annual Technical Conference, Boston, 23-25 June 2010, pp. 1-14.

6. M. Isshiki, M. Hirahara and T. Kishimoto, "FEMINITY Series Home Network System for Network Home Appliances," Toshiba Review, vol. 57, no. 10, 2002, pp. 7-10.

7. NLANR/DAST, "Iperf 2.0.3," 2008. http://iperf.sourceforge.net/

8. RHOX, "NEGiES version 1.57," 2005. http://hp.vector.co.jp/authors/VA036210/

9. Japan Communications Inc., "B-Mobile3G," 2009. http://www.bmobile.ne.jp/english/3g.html

10. S. Ishida, M. Suzuki, T. Morito and H. Morikawa, "A Multi-Step Wake-Up Scheme for Low-Power-Listening Wireless Communication System," Technical Report of IEICE, Information Network, vol. 107, no. 525, 2008, pp. 355-360.

11. A. Russel, "Comet: Low Latency Data for the Browser," 2006. http://infrequently.org/2006/03/comet-low-latency-data-for-the-browser/

12. ORIMO inc., "ORIMO Mobile Research, a Study on Smart Phones," 2010. http://www.orimo-r.co.jp/upload_2/gif/phone0601.pdf

Index